FUNDAMENTALS OF ENGINEERING
ELECTRICAL DISCIPLINE
SAMPLE QUESTIONS & SOLUTIONS

FUNDAMENTALS OF ENGINEERING

ELECTRICAL DISCIPLINE
SAMPLE QUESTIONS & SOLUTIONS

Published by the
National Council of Examiners for Engineering and Surveying®
280 Seneca Creek Road, Clemson, SC 29631 800-250-3196 www.ncees.org

ISBN 1-932613-25-0

Printed in the United States of America

TABLE OF CONTENTS

UPDATES TO EXAMINATION INFORMATION

For current exam specifications, errata for this book, a list of calculators that may be used at the exam, guidelines for requesting special accommodations, and other information about exams, visit the NCEES Web site at www.ncees.org.

INTRODUCTION

One of the functions of the National Council of Examiners for Engineering and Surveying (NCEES) is to develop examinations that are taken by candidates for licensure as professional engineers. NCEES has prepared this handbook to assist candidates who are preparing for the Fundamentals of Engineering (FE) examination in general engineering. NCEES is an organization established to assist and support the licensing boards that exist in all states and U.S. territories. NCEES provides these boards with uniform examinations that are valid measures of minimum competency related to the practice of engineering.

To develop reliable and valid examinations, NCEES employs procedures using the guidelines established in the *Standards for Educational and Psychological Testing* published by the American Psychological Association. These procedures are intended to maximize the fairness and quality of the examinations. To ensure that the procedures are followed, NCEES uses experienced testing specialists possessing the necessary expertise to guide the development of examinations using current testing techniques.

The examinations are prepared by committees composed of professional engineers from throughout the nation. These engineers supply the content expertise that is essential in developing examinations that are valid measures of minimum competency.

LICENSURE: AN IMPORTANT DECISION

One of the most important decisions you can make early in your engineering career is to place yourself on a professional course and plan to become licensed as a professional engineer (P.E.). The licensure of professional engineers is important to the public because of the significant role engineering plays in society. The profession regulates itself, through the licensing boards, by setting high standards for professional engineers. These high standards help protect the public by requiring that professional engineers demonstrate their competence to practice in a manner that will safeguard the public's safety and welfare.

The first step on the path to licensure as a P.E. is to take and pass the FE examination. If you are a student or a recent college graduate, you are well advised to take this step while coursework is still fresh in your mind. After passing the examination, your state board will designate you as an engineer intern (EI). In the past, the term "engineer-in-training" (EIT) has been used to recognize this step in your career.

To continue the licensure process, typically you must complete 4 years of progressive and verifiable experience that is acceptable to your licensing board. Some boards require that this experience be gained under the supervision of a professional engineer. Again, because of variations from state to state, you should contact your licensing board for information to ensure that you are on track to meet their requirements. Having met these requirements, you will be granted permission to take the Principles and Practice of Engineering (PE) examination. After you pass the PE examination, you may become licensed as a professional engineer and use the distinguished P.E. designation.

Licensing Requirements

Eligibility

The primary purpose of licensure is to protect the public by evaluating the qualifications of candidates seeking licensure. While examinations offer one means of measuring the competency levels of candidates, most licensing boards also screen candidates based on education and experience requirements. Because these requirements vary between boards, it would be wise to contact the appropriate board. Board addresses and telephone numbers may be obtained by visiting our Web site at www.ncees.org or by calling 800-250-3196.

Application Procedures and Deadlines

Application procedures for the examination and instructional information are available from individual boards. Requirements and fees vary among the boards, and applicants are responsible for contacting their board office. Sufficient time must be allotted to complete the application process and assemble required data.

Description of Examinations

Examination Schedule

The NCEES FE examination is offered to the boards in the spring and fall of each year. Dates of future administrations are published on the NCEES Web site at www.ncees.org. You should contact your board for specific locations of exam sites.

Examination Content

The purpose of the FE examination is to determine if the examinee has an adequate understanding of basic science, mathematics, engineering science, engineering economics, and discipline-specific subjects normally covered in coursework taken in the last 2 years of an engineering bachelor degree program. The examination identifies those applicants who have demonstrated an acceptable level of competence in these subjects.

The 8-hour FE examination is a no-choice examination in a multiple-choice format. The examination is administered in two 4-hour sessions. The morning session contains 120 questions, and the afternoon session contains 60 questions. Each question has four answer options. The examination specifications presented in this book give details of the subjects covered on the examinations.

Numerical questions are posed mostly in metric units, normally International System of Units (SI). However, because some subjects are typically taught in U.S. customary units (in.-lb) only, questions on the examination dealing with these subject areas are posed in U.S. customary units.

The FE examination is a closed-book examination. However, since engineers rely heavily on reference materials, you will be given a copy of the NCEES *FE Supplied-Reference Handbook* at the examination site. The *Handbook* contains formulas and data that examinees cannot reasonably be expected to commit to memory. The *Handbook* does not contain all the information required to answer every question on the examination. For example, basic theories, formulas, and definitions that examinees are expected to know have not been included. To familiarize yourself with the content of the *Handbook* before the examination, visit the NCEES Web site at www.ncees.org to view and download a free copy of the *Handbook*. You may also order, at minimal cost, a hard copy of the *Handbook* on the NCEES Web site or by calling NCEES Customer Service at 800-250-3196. You will not be allowed to take your copy of the *Handbook* into the examination; you must use the copy provided to you by the proctor in the examination room.

A sample examination is presented in this book. By illustrating the general content of the subject areas and formats, the questions should be helpful in preparing for the examination. Solutions are presented for all the questions. The solution presented may not be the only way to solve the question. The intent is to demonstrate the typical effort required to solve each question.

No representation is made or intended as to future examination questions, content, or subject matter.

Examination Preparation and Review

Examination development and review workshops are conducted at least twice annually by standing committees of the NCEES. Additionally, workshops are held as required to supplement the bank of questions available. The content and format of the questions are reviewed by the committee members for compliance with the specifications and to ensure the quality and fairness of the examination. These engineers are selected with the objective that they be representative of the profession in terms of geography, ethnic background, gender, and area of practice.

SCORING OF THE EXAMINATION

Both sessions of the FE examination are worth the same total number of points, and questions are weighted at one point for each morning question and two points for each afternoon question. Within each session, every question has equal weight. Your final score on the examination is arrived at by the summation of the numbers of points obtained in each session.

The score required to pass the examination varies between administrations of the examination. This acknowledges the fact that variations in difficulty may exist between different versions of the examination. The individual passing scores are set to reflect a minimum level of competency consistent with the purpose of licensing examinations. This procedure assures candidates that they have the same chance of passing the examination even though the difficulty of an examination may vary from one administration to another.

To accomplish the setting of fair passing scores that reflect the standard of minimum competency, NCEES conducts passing score studies on a regular basis. At these studies, a representative panel of engineers familiar with the candidate population uses a criterion-referenced procedure to set the passing score for the examination. The panel discusses the concept of minimum competence and develops a written standard that clearly articulates what skills and knowledge are required of engineers if they are to practice in a manner that protects the public health, safety, and welfare.

NCEES does not use fixed-percentage passing scores such as 70% or 75% because these fail to take into account the difficulty levels of different questions that make up the examinations. Similarly, NCEES avoids "grading on the curve" because licensure is designed to ensure that practitioners possess enough knowledge to perform professional activities in a manner that protects the public welfare. The key issue is whether an individual candidate is competent to practice and not whether the candidate is better or worse than other candidates.

The passing score can vary from one administration of the examination to another to reflect differences in difficulty levels of the examinations. However, the passing score is always based on the standard of minimum competency. To avoid confusion that might arise from fluctuations in the passing score, scores are converted to a standard scale which adopts 70 as the passing score. This technique of converting to a standard scale is commonly employed by testing specialists.

EXAMINATION PROCEDURES AND INSTRUCTIONS

Visit the NCEES Web site for current information about exam procedures and instructions.

Examination Materials

Before the morning and afternoon sessions, proctors will distribute examination booklets containing an answer sheet. You should not open the examination booklet until you are instructed to do so by the proctor. Read the instructions and information given on the front and back covers, and listen carefully to all the instructions the proctor reads.

The answer sheets for the multiple-choice questions are machine scored. For proper scoring, the answer spaces should be blackened completely. Since April 2002, NCEES has provided mechanical pencils with 0.7-mm HB lead to be used in the examination. You are not permitted to use any other writing instrument. If you decide to change an answer, you must erase the first answer completely. Incomplete erasures and stray marks may be read as intended answers. One side of the answer sheet is used to collect identification and biographical data. Proctors will guide you through the process of completing this portion of the answer sheet prior to taking the test. This process will take approximately 15 minutes.

Starting and Completing the Examination

You are not to open the examination booklet until instructed to do so by your proctor. If you complete the examination with more than 15 minutes remaining, you are free to leave after returning all examination materials to the proctor. Within 15 minutes of the end of the examination, you are required to remain until the end to avoid disruption to those still working and to permit orderly collection of all examination materials. Regardless of when you complete the examination, you are responsible for returning the numbered examination booklet assigned to you. Cooperate with the proctors collecting the examination materials. Nobody will be allowed to leave until the proctor has verified that all materials have been collected.

Calculators

Beginning with the April 2004 exam administration, the NCEES has strictly prohibited certain calculators from exam sites. Devices having a QWERTY keypad arrangement similar to a typewriter or keyboard are not permitted. Devices not permitted include but are not limited to palmtop, laptop, handheld, and desktop computers, calculators, databanks, data collectors, and organizers. The NCEES Web site (www.ncees.org) gives specific details on calculators.

Special Accommodations

The NCEES document *Guidelines for Requesting Religious and ADA Accommodations* explains the requirements for taking an NCEES exam with special testing accommodations. Candidates who wish to request special testing accommodations should refer to the NCEES Web site (www.ncees.org) under the "Exams" heading to find this document, along with frequently asked questions and forms for making the requests. To allow adequate evaluation time, NCEES must receive requests no later than 60 days prior to the exam administration.

EXAM SPECIFICATIONS FOR THE MORNING SESSION

NATIONAL COUNCIL OF EXAMINERS FOR ENGINEERING AND SURVEYING

Fundamentals of Engineering (FE) Examination

Effective October 2005

- The FE examination is an 8-hour supplied-reference examination: 120 questions in the 4-hour morning session and 60 questions in the 4-hour afternoon session.
- The afternoon session is administered in the following seven modules—Chemical, Civil, Electrical, Environmental, Industrial, Mechanical, and Other/General engineering.
- Examinees work all questions in the morning session and all questions in the afternoon module they have chosen.

MORNING SESSION: (120 questions in 12 topic areas)

Topic Area	Approximate Percentage of Test Content
I. Mathematics	**15%**
A. Analytic geometry	
B. Integral calculus	
C. Matrix operations	
D. Roots of equations	
E. Vector analysis	
F. Differential equations	
G. Differential calculus	
II. Engineering Probability and Statistics	**7%**
A. Measures of central tendencies and dispersions (e.g., mean, mode, standard deviation)	
B. Probability distributions (e.g., discrete, continuous, normal, binomial)	
C. Conditional probabilities	
D. Estimation (e.g., point, confidence intervals) for a single mean	
E. Regression and curve fitting	
F. Expected value (weighted average) in decision-making	
G. Hypothesis testing	
III. Chemistry	**9%**
A. Nomenclature	
B. Oxidation and reduction	
C. Periodic table	
D. States of matter	
E. Acids and bases	
F. Equations (e.g., stoichiometry)	
G. Equilibrium	
H. Metals and nonmetals	

Topic Area	Approximate Percentage of Test Content
IV. Computers	**7%**
A. Terminology (e.g., memory types, CPU, baud rates, Internet)	
B. Spreadsheets (e.g., addresses, interpretation, "what if," copying formulas)	
C. Structured programming (e.g., assignment statements, loops and branches, function calls)	
V. Ethics and Business Practices	**7%**
A. Code of ethics (professional and technical societies)	
B. Agreements and contracts	
C. Ethical versus legal	
D. Professional liability	
E. Public protection issues (e.g., licensing boards)	
VI. Engineering Economics	**8%**
A. Discounted cash flow (e.g., equivalence, PW, equivalent annual FW, rate of return)	
B. Cost (e.g., incremental, average, sunk, estimating)	
C. Analyses (e.g., breakeven, benefit-cost)	
D. Uncertainty (e.g., expected value and risk)	
VII. Engineering Mechanics (Statics and Dynamics)	**10%**
A. Resultants of force systems	
B. Centroid of area	
C. Concurrent force systems	
D. Equilibrium of rigid bodies	
E. Frames and trusses	
F. Area moments of inertia	
G. Linear motion (e.g., force, mass, acceleration, momentum)	
H. Angular motion (e.g., torque, inertia, acceleration, momentum)	
I. Friction	
J. Mass moments of inertia	
K. Impulse and momentum applied to:	
1. particles	
2. rigid bodies	
L. Work, energy, and power as applied to:	
1. particles	
2. rigid bodies	
VIII. Strength of Materials	**7%**
A. Shear and moment diagrams	
B. Stress types (e.g., normal, shear, bending, torsion)	
C. Stress strain caused by:	
1. axial loads	
2. bending loads	
3. torsion	
4. shear	

Topic Area	Approximate Percentage of Test Content
D. Deformations (e.g., axial, bending, torsion)	
E. Combined stresses	
F. Columns	
G. Indeterminant analysis	
H. Plastic versus elastic deformation	
IX. Material Properties	**7%**
A. Properties	
1. chemical	
2. electrical	
3. mechanical	
4. physical	
B. Corrosion mechanisms and control	
C. Materials	
1. engineered materials	
2. ferrous metals	
3. nonferrous metals	
X. Fluid Mechanics	**7%**
A. Flow measurement	
B. Fluid properties	
C. Fluid statics	
D. Energy, impulse, and momentum equations	
E. Pipe and other internal flow	
XI. Electricity and Magnetism	**9%**
A. Charge, energy, current, voltage, power	
B. Work done in moving a charge in an electric field (relationship between voltage and work)	
C. Force between charges	
D. Current and voltage laws (Kirchhoff, Ohm)	
E. Equivalent circuits (series, parallel)	
F. Capacitance and inductance	
G. Reactance and impedance, susceptance and admittance	
H. AC circuits	
I. Basic complex algebra	
XII. Thermodynamics	**7%**
A. Thermodynamic laws (e.g., 1st Law, 2nd Law)	
B. Energy, heat, and work	
C. Availability and reversibility	
D. Cycles	
E. Ideal gases	
F. Mixture of gases	
G. Phase changes	
H. Heat transfer	
I. Properties of:	
1. enthalpy	
2. entropy	

MORNING SAMPLE QUESTIONS

NOTE: THESE QUESTIONS REPRESENT ONLY HALF THE NUMBER OF QUESTIONS THAT APPEAR ON THE ACTUAL EXAMINATION.

MORNING SAMPLE QUESTIONS

1. If the functional form of a curve is known, differentiation can be used to determine all of the following **EXCEPT** the:

(A) concavity of the curve

(B) location of inflection points on the curve

(C) number of inflection points on the curve

(D) area under the curve between certain bounds

2. Which of the following is the general solution to the differential equation and boundary condition shown below?

$$\frac{dy}{dt} + 5y = 0;\ y(0) = 1$$

(A) e^{5t}

(B) e^{-5t}

(C) $e^{\sqrt{-5t}}$

(D) $5e^{-5t}$

GO ON TO THE NEXT PAGE

3. If D is the differential operator, then the general solution to $(D + 2)^2y = 0$ is:

(A) C_1e^{-4x}

(B) C_1e^{-2x}

(C) $e^{-4x}(C_1 + C_2x)$

(D) $e^{-2x}(C_1 + C_2x)$

4. A particle traveled in a straight line in such a way that its distance S from a given point on that line after time t was $S = 20t^3 - t^4$. The rate of change of acceleration at time $t = 2$ is:

(A) 72
(B) 144
(C) 192
(D) 208

GO ON TO THE NEXT PAGE

5. Which of the following is a unit vector perpendicular to the plane determined by the vectors $\mathbf{A} = 2\mathbf{i} + 4\mathbf{j}$ and $\mathbf{B} = \mathbf{i} + \mathbf{j} - \mathbf{k}$?

(A) $-2\mathbf{i} + \mathbf{j} - \mathbf{k}$

(B) $\frac{1}{\sqrt{5}}(\mathbf{i} + 2\mathbf{j})$

(C) $\frac{1}{\sqrt{6}}\left(-2\mathbf{i} + \mathbf{j} - \mathbf{k}\right)$

(D) $\frac{1}{\sqrt{6}}\left(-2\mathbf{i} - \mathbf{j} - \mathbf{k}\right)$

6. If f' denotes the derivative of a function of $y = f(x)$, then $f'(x)$ is defined by:

(A) $\lim\limits_{\Delta y \to 0} \frac{\Delta x}{\Delta y}$

(B) $\lim\limits_{\Delta y \to 0} \frac{\Delta y}{\Delta x}$

(C) $\lim\limits_{\Delta x \to 0} \frac{f(x + \Delta x) - f(x)}{\Delta x}$

(D) $\lim\limits_{\Delta y \to 0} \frac{f(x - \Delta x) + f(x)}{\Delta y}$

GO ON TO THE NEXT PAGE

7. What is the area of the region in the first quadrant that is bounded by the line $y = 1$, the curve $x = y^{3/2}$, and the y-axis?

(A) 2/5
(B) 3/5
(C) 2/3
(D) 1

8. Three lines are defined by the three equations:

$$\begin{aligned} x + y &= 0 \\ x - y &= 0 \\ 2x + y &= 1 \end{aligned}$$

The three lines form a triangle with vertices at:

(A) $(0, 0), \left(\frac{1}{3}, \frac{1}{3}\right), (1, -1)$

(B) $(0, 0), \left(\frac{2}{3}, \frac{2}{3}\right), (-1, -1)$

(C) $(1, 1), (1, -1), (2, 1)$

(D) $(1, 1), (3, -3), (-2, -1)$

GO ON TO THE NEXT PAGE

9. The value of the integral $\int_0^{\pi} 10 \sin x dx$ is:

(A) –10
(B) 0
(C) 10
(D) 20

10. You wish to estimate the mean M of a population from a sample of size n drawn from the population. For the sample, the mean is x and the standard deviation is s. The probable accuracy of the estimate improves with an increase in:

(A) M
(B) n
(C) s
(D) $M + s$

GO ON TO THE NEXT PAGE

11. A bag contains 100 balls numbered from 1 to 100. One ball is removed. What is the probability that the number on this ball is odd or greater than 80?

(A) 0.2
(B) 0.5
(C) 0.6
(D) 0.8

12. The standard deviation of the population of the three values 1, 4, and 7 is:

(A) $\sqrt{3}$

(B) $\sqrt{6}$

(C) 4

(D) 6

GO ON TO THE NEXT PAGE

13. Suppose the lengths of telephone calls form a normal distribution with a mean length of 8.0 min and a standard deviation of 2.5 min. The probability that a telephone call selected at random will last more than 15.5 min is most nearly:

(A) 0.0013
(B) 0.0026
(C) 0.2600
(D) 0.9987

14. The volume (L) of 1 mol of H_2O at 546 K and 1.00 atm pressure is most nearly:

(A) 11.2
(B) 14.9
(C) 22.4
(D) 44.8

GO ON TO THE NEXT PAGE

15. Consider the equation:

$$As_2O_3 + 3\,C \rightarrow 3\,CO + 2\,As$$

Atomic weights may be taken as 75 for arsenic, 16 for oxygen, and 12 for carbon. According to the equation above, the reaction of 1 standard gram-mole of As_2O_3 with carbon will result in the formation of:

(A) 1 gram-mole of As

(B) 28 grams of CO

(C) 150 grams of As

(D) a greater amount by weight of CO than of As

16. If 60 mL of NaOH solution neutralizes 40 mL of 0.50 M H_2SO_4, the concentration of the NaOH solution is most nearly:

(A) 0.80 M
(B) 0.67 M
(C) 0.45 M
(D) 0.33 M

GO ON TO THE NEXT PAGE

17. The atomic weights of sodium, oxygen, and hydrogen are 23, 16 and 1, respectively. To neutralize 4 grams of NaOH dissolved in 1 L of water requires 1 L of:

(A) 0.001 normal HCl solution

(B) 0.01 normal HCl solution

(C) 0.1 normal HCl solution

(D) 1.0 normal HCl solution

18. Consider the following equation:

$$K = \frac{[C]^2[D]^2}{[A]^4[B]}$$

The equation above is the formulation of the chemical equilibrium constant equation for which of the following reactions?

(A) $C_2 + D_2 \leftrightarrow A_4 + B$

(B) $4A + B \leftrightarrow 2C + 2D$

(C) $4C + 2D \leftrightarrow 2A + B$

(D) $A_4 + B \leftrightarrow C_2 + D_2$

GO ON TO THE NEXT PAGE

19. The flowchart for a computer program contains the following segment:

```
   VAR = 0
┌→ IF VAR < 5 THEN VAR = VAR + 2
│  OTHERWISE EXIT LOOP
└─ LOOP
```

What is the value of VAR at the conclusion of this routine?

(A) 0
(B) 2
(C) 4
(D) 6

20. In a spreadsheet, the number in Cell A4 is set to 6. Then A5 is set to A4 + A4. This formula is copied into Cells A6 and A7. The number shown in Cell A7 is most nearly:

(A) 12
(B) 24
(C) 36
(D) 216

GO ON TO THE NEXT PAGE

21. Consider the following program segment:

```
INPUT Z, N
S = 1
T = 1
FOR K = 1 TO N
T = T*Z/K
S = S + T
NEXT K
```

This segment calculates the sum:

(A) $S = 1 + ZT + 2\,ZT + 3\,ZT + \ldots + N\,ZT$

(B) $S = 1 + ZT + \frac{1}{2}ZT + \frac{1}{3}ZT + \ldots + \left(\frac{1}{N}\right)ZT$

(C) $S = 1 + \frac{Z}{1} + \frac{2Z}{2} + \frac{3Z}{3} + \ldots + \left(\frac{NZ}{N}\right)$

(D) $S = 1 + \frac{Z}{1!} + \frac{Z^2}{2!} + \frac{Z^3}{3!} + \ldots + \left(\frac{Z^N}{N!}\right)$

22. In a spreadsheet, Row 1 has the numbers 2, 4, 6, 8, 10, ... , 20 in Columns A–J, and Row 2 has the numbers 1, 3, 5, 7, 9, ... , 19 in the same columns. All other cells are zero except for Cell D3, which contains the formula: D1 + D$1*D2. This formula has been copied into cells D4 and D5. The number that appears in cell D4 is most nearly:

(A) 3
(B) 64
(C) 519
(D) 4,216

GO ON TO THE NEXT PAGE

23. An engineer testifying as an expert witness in a product liability case should:

(A) answer as briefly as possible only those questions posed by the attorneys

(B) provide a complete and objective analysis within his or her area of competence

(C) provide an evaluation of the character of the defendant

(D) provide information on the professional background of the defendant

24. A professional engineer, originally licensed 30 years ago, is asked to evaluate a newly developed computerized control system for a public transportation system. The engineer may accept this project if:

(A) he or she is competent in the area of modern control systems

(B) his or her professional engineering license has not lapsed

(C) his or her original area of specialization was in transportation systems

(D) he or she has regularly attended meetings of a professional engineering society

25. You and your design group are competing for a multidisciplinary concept project. Your firm is the lead group in the design professional consortium formed to compete for the project. Your consortium has been selected as the first to enter fee negotiations with the project owner. During the negotiations, the amount you have to cut from your fee to be awarded the contract will require dropping one of the consortium members whose staff has special capabilities not available from the staff of the remaining consortium members. Can your remaining consortium ethically accept the contract?

(A) No, because an engineer may not accept a contract to coordinate a project with other professional firms providing capabilities and services that must be provided by hired consultants.

(B) Yes, if your remaining consortium members hire a few new lower-cost employees to do the special work that would have been provided by the consortium member that has been dropped.

(C) No, not if the owner is left with the impression that the consortium is still fully qualified to perform all the required tasks.

(D) Yes, if in accepting an assignment to coordinate the project, a single person will sign and seal all the documents for the entire work of the consortium.

26. You have an on-site job interview to follow up on an on-campus interview with Company A. Just before you fly to the interview, you get a call from Company B asking you to come for an on-site interview at their offices in the same city. When you inform them of your interview with Company A, they suggest you stop in after that. Company A has already paid for your airfare and, at the conclusion of your interview with them, issues you reimbursement forms for the balance of your trip expenses with instructions to file for all your trip expenses. When you inform them of your added interview stop at Company B, they tell you to go ahead and charge the entire cost of the trip to Company A. You interview with Company B, and at the conclusion, they give you travel reimbursement forms with instructions to file for all your trip expenses. When you inform them of the instructions of Company A, they tell you that the only expenses requiring receipts are airfare and hotel rooms, so you should still file for all the other expenses with them even if Company A is paying for it because students always need a little spending money. What should you do?

(A) Try to divide the expenses between both firms as best you can.

(B) Do as both recruiting officers told you. It is their money and their travel policies.

(C) File for travel expenses with only one firm.

(D) Tell all your classmates to sign up to interview with these firms for the trips.

GO ON TO THE NEXT PAGE

27. A company can manufacture a product using hand tools. Costs will be $1,000 for tools and a $1.50 manufacturing cost per unit. As an alternative, an automated system will cost $15,000 with a $0.50 manufacturing cost per unit. With an anticipated annual volume of 5,000 units and neglecting interest, the breakeven point (years) is most nearly:

(A) 2.8
(B) 3.6
(C) 15.0
(D) never

28. A printer costs $900. Its salvage value after 5 years is $300. Annual maintenance is $50. If the interest rate is 8%, the equivalent uniform annual cost is most nearly:

(A) $224
(B) $300
(C) $327
(D) $350

29. The need for a large-capacity water supply system is forecast to occur 4 years from now. At that time, the system required is estimated to cost $40,000. If an account earns 12% per year compounded annually, the amount that must be placed in the account at the end of each year in order to accumulate the necessary purchase price is most nearly:

(A) $8,000
(B) $8,370
(C) $9,000
(D) $10,000

30. A project has the estimated cash flows shown below.

Year End	0	1	2	3	4
Cash Flow	–$1,100	–$400	+$1,000	+$1,000	+$1,000

Using an interest rate of 12% per year compounded annually, the annual worth of the project is most nearly:

(A) $450
(B) $361
(C) $320
(D) $226

31. You must choose between four pieces of comparable equipment based on the cash flows given below. All four pieces have a life of 8 years.

Parameter	Equipment			
	A	B	C	D
First cost	$25,000	$35,000	$20,000	$40,000
Annual costs	$8,000	$6,000	$9,000	$5,000
Salvage value	$2,500	$3,500	$2,000	$4,000

The discount rate is 12%. Ignore taxes. The most preferable top two projects and the difference between their present worth values are most nearly:

(A) A and C, $170
(B) B and D, $170
(C) A and C, $234
(D) B and D, $234

32. Referring to the figure below, the coefficient of static friction between the block and the inclined plane is 0.25. The block is in equilibrium.

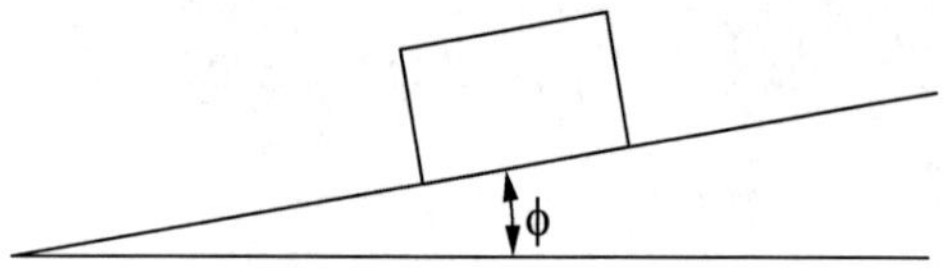

As the inclined plane is raised, the block will begin to slide when:

(A) $\sin \phi = 1.0$

(B) $\cos \phi = 1.0$

(C) $\cos \phi = 0.25$

(D) $\tan \phi = 0.25$

33. A cylinder weighing 120 N rests between two frictionless walls as shown in the following figure.

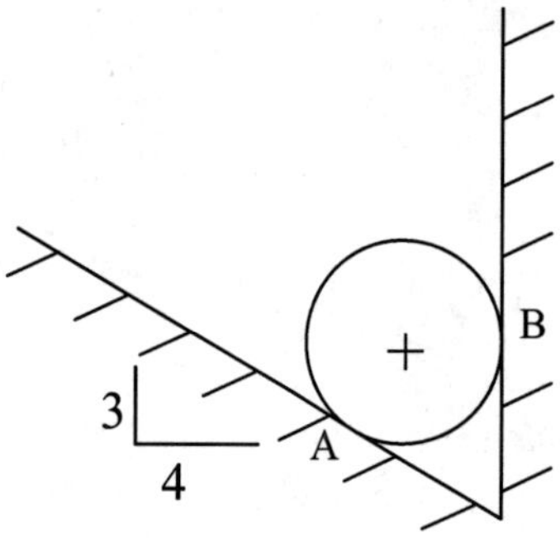

The wall reaction (N) at Point A is most nearly:

(A) 96
(B) 139
(C) 150
(D) 200

34. Three forces act as shown below.

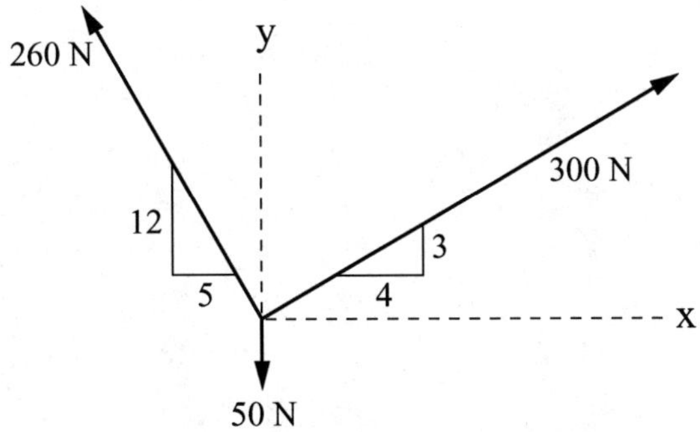

The magnitude of the resultant of the three forces (N) is most nearly:

(A) 140
(B) 191
(C) 370
(D) 396

35. In the figure below, Block A weighs 50 N, Block B weighs 80 N, and Block C weighs 100 N. The coefficient of friction at all surfaces is 0.30.

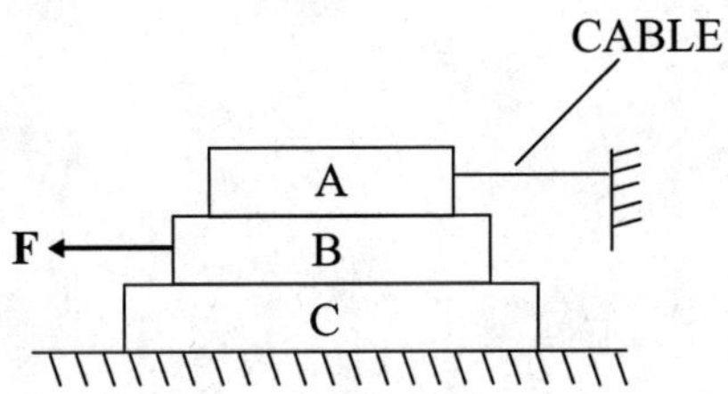

The maximum force **F** (N) that can be applied to Block B without disturbing equilibrium is most nearly:

(A) 15
(B) 54
(C) 69
(D) 84

36. The moment of force **F** (N•m) shown below with respect to Point p is most nearly:

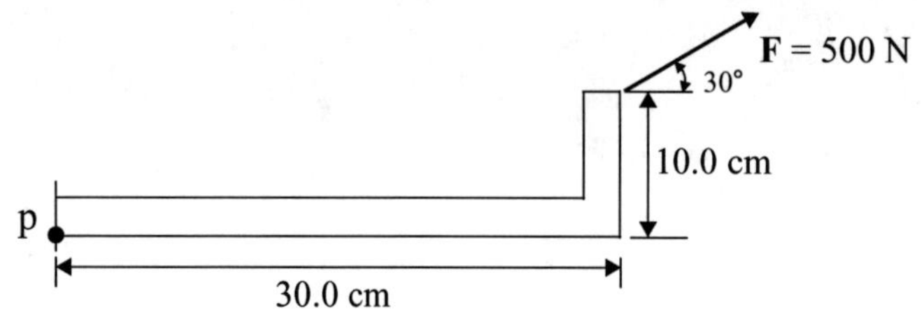

(A) 31.7 ccw
(B) 31.7 cw
(C) 43.3 cw
(D) 43.3 ccw

37. The figure below shows a simple truss.

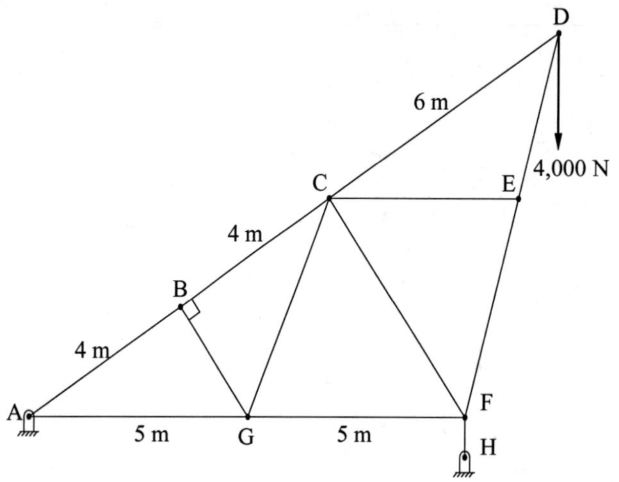

The zero-force members in the truss are:

(A) BG, CG, CF, CE
(B) BG, CE
(C) CF
(D) CG, CF

38. The beam shown below is known as a:

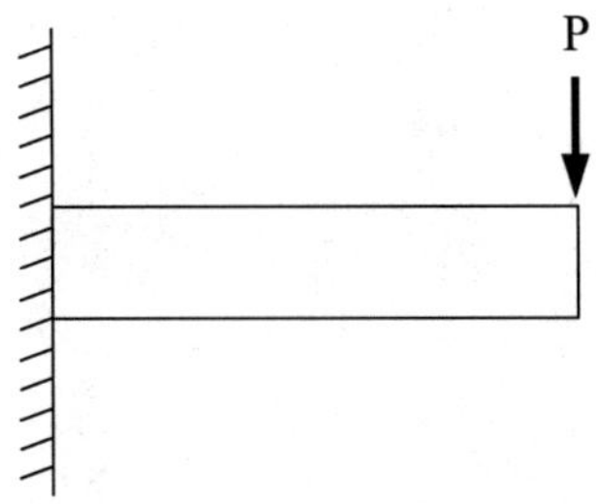

(A) cantilever beam
(B) statically indeterminate beam
(C) simply supported beam
(D) continuously loaded beam

39. The shear diagram for a particular beam is shown below. All lines in the diagram are straight. The bending moment at each end of the beam is zero, and there are no concentrated couples along the beam.

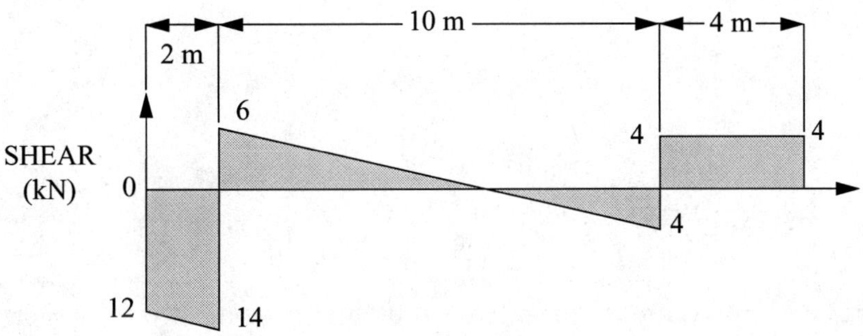

The maximum magnitude of the bending moment (kN·m) in the beam is most nearly:

(A) 8
(B) 16
(C) 18
(D) 26

40. The piston of a steam engine is 50 cm in diameter, and the maximum steam gage pressure is 1.4 MPa. If the design stress for the piston rod is 68 MPa, its cross-sectional area (m^2) should be most nearly:

(A) 40.4×10^{-4}
(B) 98.8×10^{-4}
(C) 228.0×10^{-4}
(D) 323.0×10^{-4}

41. A shaft of wood is to be used in a certain process. If the allowable shearing stress parallel to the grain of the wood is 840 kN/m^2, the torque (N•m) transmitted by a 200-mm-diameter shaft with the grain parallel to the neutral axis is most nearly:

(A) 500
(B) 1,200
(C) 1,320
(D) 1,500

42. The Euler formula for columns deals with:

(A) relatively short columns
(B) shear stress
(C) tensile stress
(D) elastic buckling

GO ON TO THE NEXT PAGE

43. The mechanical deformation of a material above its recrystallization temperature is commonly known as:

(A) hot working

(B) strain aging

(C) grain growth

(D) cold working

44. In general, a metal with high hardness will also have:

(A) good formability

(B) high impact strength

(C) high electrical conductivity

(D) high yield strength

45. Glass is said to be an amorphous material. This means that it:

(A) has a high melting point

(B) is a supercooled vapor

(C) has large cubic crystals

(D) has no apparent crystal structure

GO ON TO THE NEXT PAGE

46. If an aluminum crimp connector were used to connect a copper wire to a battery, what would you expect to happen?

(A) The copper wire only will corrode.

(B) The aluminum connector only will corrode.

(C) Both will corrode.

(D) Nothing

47. The rectangular homogeneous gate shown below is 3.00 m high × 1.00 m wide and has a frictionless hinge at the bottom. If the fluid on the left side of the gate has a density of 1,600 kg/m^3, the magnitude of the force **F** (kN) required to keep the gate closed is most nearly:

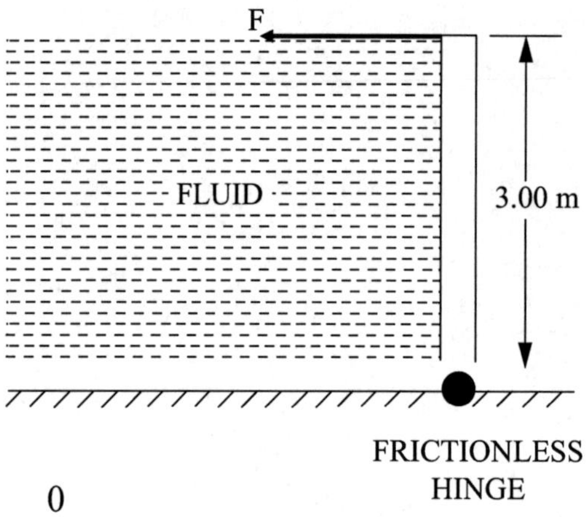

(A) 0
(B) 22
(C) 24
(D) 220

GO ON TO THE NEXT PAGE

48. Which of the following statements is true of viscosity?

(A) It is the ratio of inertial to viscous force.

(B) It always has a large effect on the value of the friction factor.

(C) It is the ratio of the shear stress to the rate of shear deformation.

(D) It is usually low when turbulent forces predominate.

49. A horizontal jet of water (density = 1,000 kg/m^3) is deflected perpendicularly to the original jet stream by a plate as shown below.

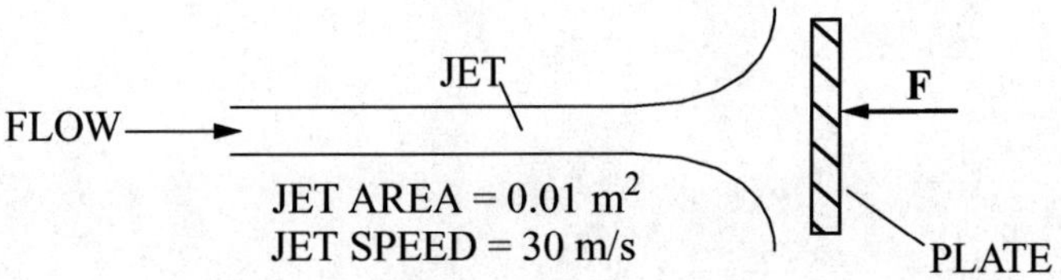

The magnitude of force **F** (kN) required to hold the plate in place is most nearly:

(A) 4.5
(B) 9.0
(C) 45.0
(D) 90.0

50. Which of the following statements about flow through an insulated valve is most accurate?

(A) The enthalpy rises.

(B) The upstream and downstream enthalpies are equal.

(C) Temperature increases sharply.

(D) Pressure increases sharply.

51. The pitot tube shown below is placed at a point where the velocity is 2.0 m/s. The specific gravity of the fluid is 2.0, and the upper portion of the manometer contains air. The reading h (m) on the manometer is most nearly:

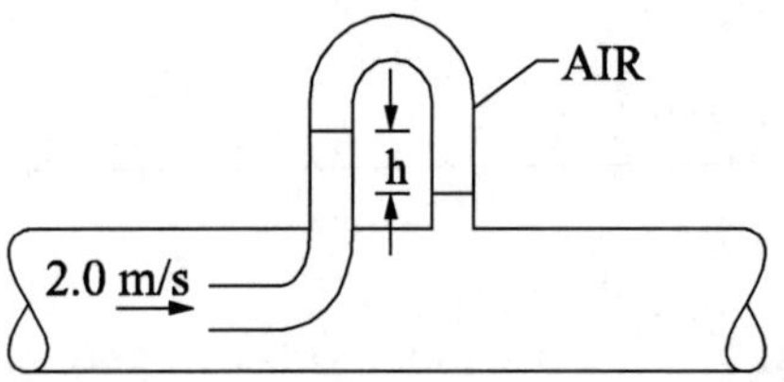

(A) 20.0
(B) 10.0
(C) 0.40
(D) 0.20

52. If the complex power is 1,500 VA with a power factor of 0.866 lagging, the reactive power (VAR) is most nearly:

(A) 0
(B) 750
(C) 1,300
(D) 1,500

53. Series-connected circuit elements are shown in the figure below.

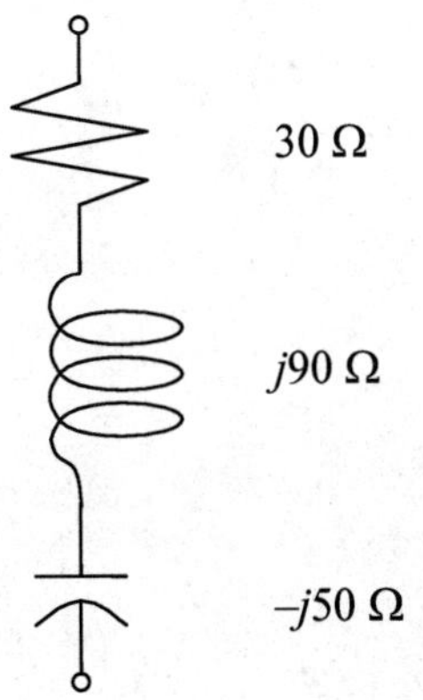

Which of the following impedance diagrams is correct according to conventional notation?

(A)

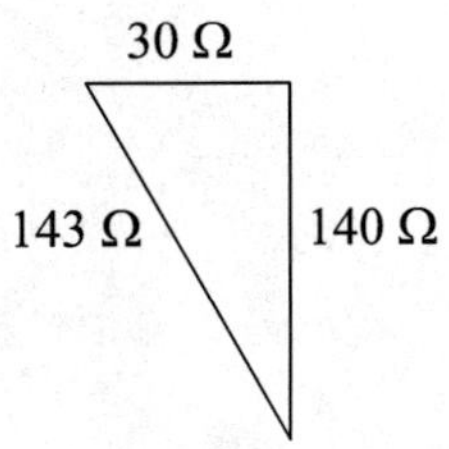

(B)

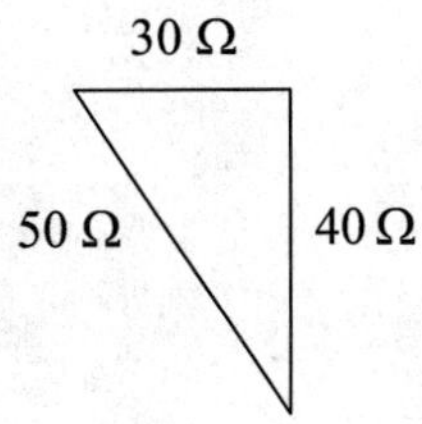

(C)

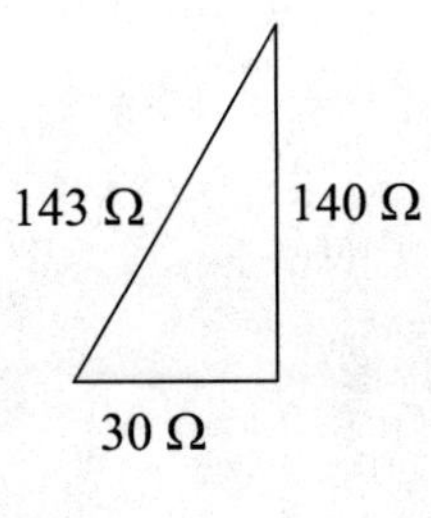

(D)

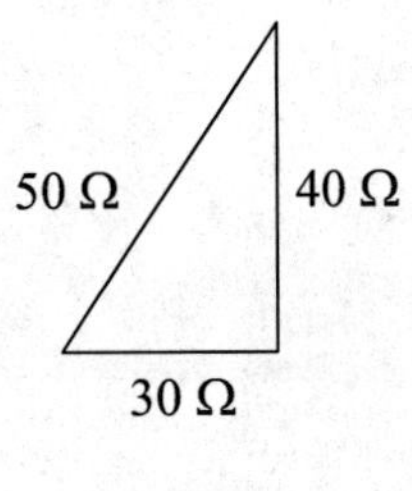

GO ON TO THE NEXT PAGE

54. A 10-μF capacitor has been charged to a potential of 150 V. A resistor of 25 Ω is then connected across the capacitor through a switch. When the switch is closed for ten time constants, the total energy (joules) dissipated by the resistor is most nearly:

(A) 1.0×10^{-7}
(B) 1.1×10^{-1}
(C) 9.0×10^{1}
(D) 9.0×10^{3}

55. The connecting wires and the battery in the circuit shown below have negligible resistance.

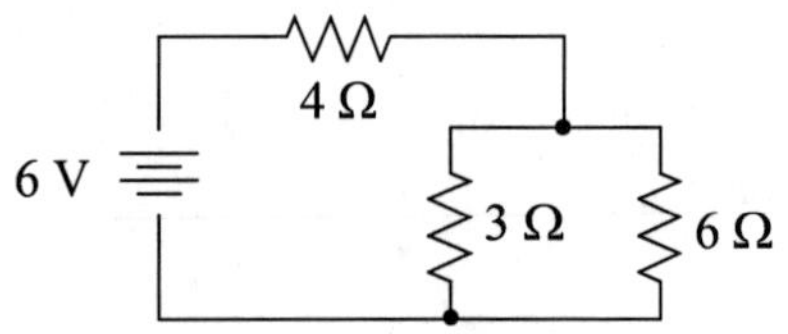

The current (amperes) through the 6-Ω resistor is most nearly:

(A) 1/3
(B) 1/2
(C) 1
(D) 3/2

GO ON TO THE NEXT PAGE

56. The term $\frac{(1-i)^2}{(1+i)^2}$, where $i = \sqrt{-1}$, is most nearly:

(A) -1
(B) $-1 + i$
(C) 0
(D) $1 + i$

57. An insulated tank contains half liquid and half vapor by volume in equilibrium. The release of a small quantity of the vapor without the addition of heat will cause:

(A) evaporation of some liquid in the tank
(B) superheating of the vapor in the tank
(C) a rise in temperature
(D) an increase in enthalpy

58. The heat transfer during an adiabatic process is:

(A) reversible
(B) irreversible
(C) dependent on temperature
(D) zero

59. An isentropic process is one which:

(A) is adiabatic but not reversible

(B) is reversible but not adiabatic

(C) is adiabatic and reversible

(D) occurs at constant pressure and temperature

60. The universal gas constant is 8.314 kJ/(kmol•K). The gas constant [(kJ/(kg•K)] of a gas having a molecular weight of 44 is most nearly:

(A) 0.19
(B) 0.38
(C) 0.55
(D) 5.3

IF YOU FINISH BEFORE TIME IS CALLED, YOU MAY WISH TO CHECK YOUR WORK ON THIS TEST.

MORNING SOLUTIONS

ANSWERS TO THE MORNING SAMPLE QUESTIONS

Detailed solutions for each question begin on the next page.

QUESTION	ANSWER	QUESTION	ANSWER
1	D	31	B
2	B	32	D
3	D	33	C
4	A	34	D
5	C	35	B
6	C	36	A
7	A	37	A
8	A	38	A
9	D	39	D
10	B	40	A
11	C	41	C
12	B	42	D
13	A	43	A
14	D	44	D
15	C	45	D
16	B	46	B
17	C	47	C
18	B	48	C
19	D	49	B
20	B	50	B
21	D	51	D
22	C	52	B
23	B	53	D
24	A	54	B
25	A	55	A
26	A	56	A
27	A	57	A
28	A	58	D
29	B	59	C
30	D	60	A

MORNING SOLUTIONS

1. The area under a curve is determined by integration, not differentiation.

THE CORRECT ANSWER IS: (D)

2. The characteristic equation for a first-order linear homogeneous differential equation is:

$$r + 5 = 0$$

which has a root at $r = -5$.

Refer to Differential Equations in the Mathematics section of the *FE Reference Handbook.* The form of the solution is then:

$$y = C\,e^{-\alpha t} \text{ where } \alpha = a \quad \text{and} \quad a = 5 \text{ for this problem}$$

C is determined from the boundary condition.

$$1 = C\,e^{-5(0)}$$
$$C = 1$$

Then, $y = e^{-5t}$

THE CORRECT ANSWER IS: (B)

3. Refer to Differential Equations in the Mathematics section of the *FE Reference Handbook.* The characteristic equation for a second-order linear homogeneous differential equation is:

$$r^2 + 2ar + b = 0$$

In this problem, $D^2 + 4D + 4 = 0$, so:

$$2a = 4 \text{ or } a = 2 \text{ and } b = 4$$

In solving the characteristic equation, it is noted that there are repeated real roots: $r_1 = r_2 = -2$

Because $a^2 = b$, the solution for this critically damped system is:

$$y(x) = (C_1 + C_2\,x)\,e^{-2x}$$

THE CORRECT ANSWER IS: (D)

4. First, the velocity is:

$$V = S' = 60t^2 - 4t^3$$

Then, the acceleration is:

$$A = S'' = 120t - 12t^2$$

Finally, the rate of change of acceleration is:

$$A' = S''' = 120 - 24t$$

When $t = 2$:

$$A' = 120 - 24(2) = 120 - 48 = 72$$

THE CORRECT ANSWER IS: (A)

5. The cross product of vectors **A** and **B** is a vector perpendicular to **A** and **B**.

$$\begin{vmatrix} \mathbf{i} & \mathbf{j} & \mathbf{k} \\ 2 & 4 & 0 \\ 1 & 1 & -1 \end{vmatrix} = \mathbf{i}(-4) - \mathbf{j}(-2-0) + \mathbf{k}(2-4) = -4\mathbf{i} + 2\mathbf{j} - 2\mathbf{k}$$

To obtain a unit vector, divide by the magnitude.

$$\text{Magnitude} = \sqrt{(-4)^2 + 2^2 + (-2)^2} = \sqrt{24} = 2\sqrt{6}$$

$$\frac{-4\mathbf{i} + 2\mathbf{j} - 2\mathbf{k}}{2\sqrt{6}} = \frac{-2\mathbf{i} + \mathbf{j} - \mathbf{k}}{\sqrt{6}}$$

THE CORRECT ANSWER IS: (C)

6. Refer to Differential Calculus in the Mathematics section of the *FE Reference Handbook.*

THE CORRECT ANSWER IS: (C)

7. Define a differential strip with length $(x - 0)$ and height dy.

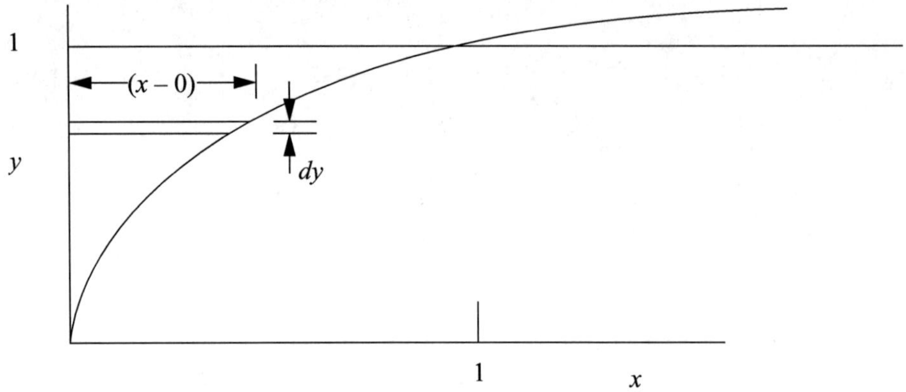

$$\int dA = \int_0^1 x\,dy = \int_0^1 y^{3/2}\,dy = \left.\frac{y^{5/2}}{5/2}\right|_0^1 = \frac{2}{5}$$

THE CORRECT ANSWER IS: (A)

8.

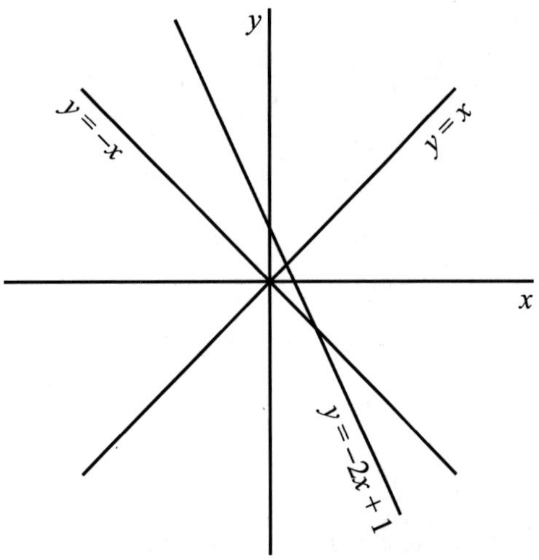

$y = -x$

$y = x$

$y = -2x + 1$

from graph one, intersection is at (0,0), so (C) and (D) are incorrect.

Also, second intersection is at (1,–1), so key has to be (A).

(0,0) (1/3, 1/3) and (1,–1)

THE CORRECT ANSWER IS: (A)

9. $$\int_0^{\pi} 10\sin x\,dx = 10\left[-\cos x\Big|_0^{\pi}\right]$$

$$= 10\left[-\cos\pi - (-\cos 0)\right]$$

$$= 10[1+1]$$

$$= 20$$

THE CORRECT ANSWER IS: (D)

10. Accuracy increases with increasing sample size.

THE CORRECT ANSWER IS: (B)

11. The key word is **OR**. What is the probability that the number is odd **OR** greater than 80? Refer to Property 2 given under Probability and Statistics in the Mathematics section of the *FE Reference Handbook.*

P(A + B) = P(A) + P(B) – P(A,B)

Event A is removing a ball with an odd number.
P(A) = 50/100 = 0.5

Event B is removing a ball with a number greater than 80.
P(B) = 20/100 = 0.2

Event A,B is removing a ball with an odd number that is greater than 80.

There are ten such balls.
P(A,B) = 10/100 = 0.1

Also P(A,B) = P(A) × P(B) = 0.5 × 0.2 = 0.1

P(A + B) = 0.5 + 0.2 – (0.5 × 0.2) = 0.6

THE CORRECT ANSWER IS: (C)

12.

x	$x-\bar{x}$	$(x-\bar{x})^2$
1	–3	9
4	0	0
7	3	9
$\Sigma = 12$		$\Sigma = 18$

$$\bar{X} = \frac{12}{3} = 4$$

$$\sigma = \sqrt{\frac{18}{3}} = \sqrt{6}$$

THE CORRECT ANSWER IS: (B)

13. $8 - 15.5 = 7.5$

$$\frac{7.5}{2.5} = 3 \text{ standard deviations}$$

From the Unit Normal Distribution Table in the Mathematics section of the *FE Reference Handbook.*

For X = 3, R(X) = 0.0013

THE CORRECT ANSWER IS: (A)

14. PV = nRT

1V = (1)(0.08206)(546)

solve for V = 44.8 L

THE CORRECT ANSWER IS: (D)

15. 2 moles of As × 75 g/mole of As = 150 g of As

THE CORRECT ANSWER IS: (C)

16. $H_2SO_4 + 2\ NaOH \rightarrow Na_2SO_4 + 2\ H_2O$

$0.5\ M\ H_2SO_4 = 1.0\ N\ H_2SO_4$

$1.0\ M\ NaOH = 1.0\ N\ NaOH$

40 mL of 1.0 N H_2SO_4 = 60 mL of x N NaOH

$40 \times 1 = 60x$

$x = 40/60 = 0.67\ N = 0.67\ M\ NaOH$

THE CORRECT ANSWER IS: (B)

17. The molecular weight of NaOH is 40 g; therefore, 4 g/L of NaOH will form 1 L of 0.1 normal NaOH solution. One liter of 0.1 normal HCl solution is required to neutralize the NaOH.

THE CORRECT ANSWER IS: (C)

18. Refer to the Chemistry section of the *FE Reference Handbook* for the equilibrium constant of a chemical reaction.

$4A + B \leftrightarrow 2C + 2D$

THE CORRECT ANSWER IS: (B)

19.

Step	VAR
1	0
2	2
3	4
4	6

EXIT LOOP

At the conclusion of the routine, VAR = 6.

THE CORRECT ANSWER IS: (D)

20.

Row	Column A	Value of A
4	6	6
5	A4 + A4	12
6	A5 + A4	18
7	A6 + A4	24

THE CORRECT ANSWER IS: (B)

21.

Step	Z	N	T	K	S
1	Z	N	.	.	.
2	Z	N	1	.	1
3	Z	N	1	1	1
.	Z	N	Z	1	1
.			Z	1	1 + Z
(NEXT K)					
			$\frac{Z^2}{2}$	2	$\frac{1+Z+Z^2}{2}$
(NEXT K)					
			$\frac{Z^3}{(2)(3)}$	3	$\frac{\frac{1+Z+Z^2}{2+Z^3}}{(2)(3)}$
(NEXT K)					
			$\frac{Z^4}{(2)(3)(4)}$	4	$\frac{\frac{\frac{1+Z+Z^2}{2+Z^3}}{(2)(3)+Z^3}}{(2)(3)(4)}$

(Z, N, T, K, S are listed under the heading "Variables".)

$\therefore$ The sequence is: $S = 1 + \frac{Z}{1!} + \frac{Z^2}{2!} + \frac{Z^3}{3!} + \frac{Z^4}{4!} + \ldots + \frac{Z^N}{N!}$

THE CORRECT ANSWER IS: (D)

22.

Rows	Columns				
	A	B	C	D	E
1	2	4	6	8	10
2	1	3	5	7	9
3				64	
4				519	
5					

D3: D1 + D$1 * D2 = 8 + 8(7) = 64

D4: D2 + D$1 * D3 = 7 + 8(64) = 519

THE CORRECT ANSWER IS: (C)

23. Refer to the NCEES *Model Rules of Professional Conduct*, Section A.4., in the Ethics section of the *FE Reference Handbook.*

THE CORRECT ANSWER IS: (B)

24. Refer to the NCEES *Model Rules of Professional Conduct*, Section B.1., in the Ethics section of the *FE Reference Handbook.*

THE CORRECT ANSWER IS: (A)

25. Refer to the NCEES *Model Rules of Professional Conduct*, Section B.3. and Section C.1., in the Ethics section of the *FE Reference Handbook.*

THE CORRECT ANSWER IS: (A)

26. Refer to the NCEES *Model Rules of Professional Conduct*, Section B.5. and Section B.6., in the Ethics section of the *FE Reference Handbook.*

THE CORRECT ANSWER IS: (A)

27. $1.50 (5,000) = $7,500

$0.50 (5,000) = $2,500

Annual savings = $7,500 – $2,500 = $5,000

Additional investment = $15,000 – $1,000 = $14,000

Payback = $14,000/$5,000 = 2.8 years

THE CORRECT ANSWER IS: (A)

28. Annual cost: = $900(A/P, 8%, 5) + $50 – $300(A/F, 8%, 5)
= $900(0.2505) + $50 – $300(0.1705)
= $225.45 + $50 – $51.15
= $224.30

THE CORRECT ANSWER IS: (A)

29.

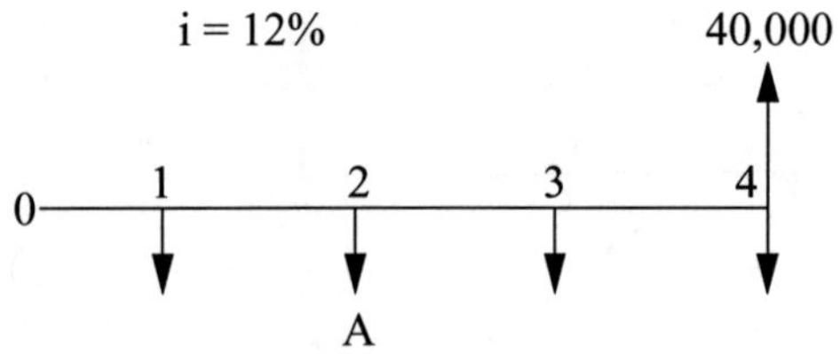

$A = F(A/F, i, n) = 40{,}000(A/F, 12\%, 4) = \$8{,}369$ per year

THE CORRECT ANSWER IS: (B)

30.

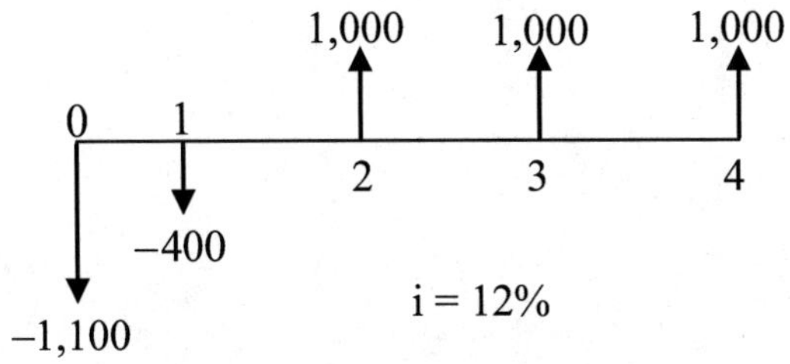

$$PW = -1{,}100 - 400\,(P/F, 12\%, 1) + 1{,}000\,(P/F, 12\%, 2)$$

$$+ 1{,}000\,(P/F, 12\%, 3) + 1{,}000\,(P/F, 12\%, 4)$$

$$= -1{,}100 - 400\,(0.8929) + 1{,}000\,(0.7972) + 1{,}000\,(0.7118) + 1{,}000\,(0.6355)$$

$$= 687.34$$

$$A = PW\,(A/P, 12\%, 4) = 687.34(0.3292)$$

$$= \$226 \text{ per year}$$

THE CORRECT ANSWER IS: (D)

31. The easiest way to solve this problem is to look at the present worth of each alternative.

The present worth values are all given by

PW = First Cost + Annual Cost × (P/A, 12%, 8) – Salvage Value × (P/F, 12%, 8)
= First Cost + Annual Cost × 4.9676 – Salvage Value × 0.4039

Then PW(A) = \$63,731
PW(B) = \$63,392
PW(C) = \$63,901
PW(D) = \$63,222

The cash flows are all costs, so the most preferable two projects, those with the lowest present worth costs, are B and D, and the difference between them is \$170.

THE CORRECT ANSWER IS: (B)

32. Normal to the plane:

$$\Sigma F_n = 0: N - mg\cos\phi = 0 \rightarrow N = mg\cos\phi$$

Tangent to the plane:

$$\Sigma F_t = 0: -mg\sin\phi + \mu N = 0$$

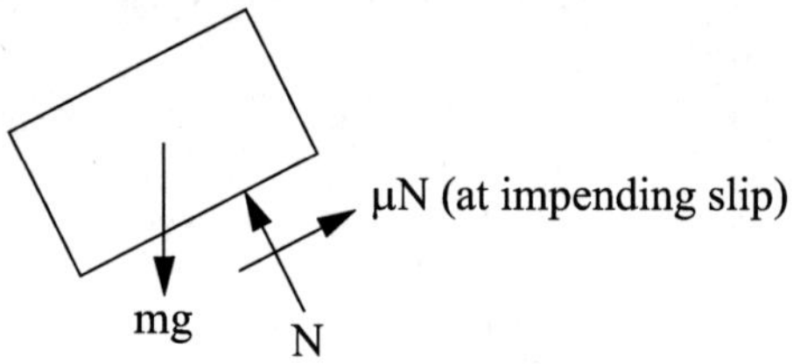

$$\therefore -mg\sin\phi + \mu mg\cos\phi = 0$$

$$\frac{\sin\phi}{\cos\phi} = \tan\phi = \mu$$

$$\tan\phi = 0.25$$

THE CORRECT ANSWER IS: (D)

33. $\Sigma F_y = 0 = -120 + \frac{4}{5}A$

$A = 150$ N

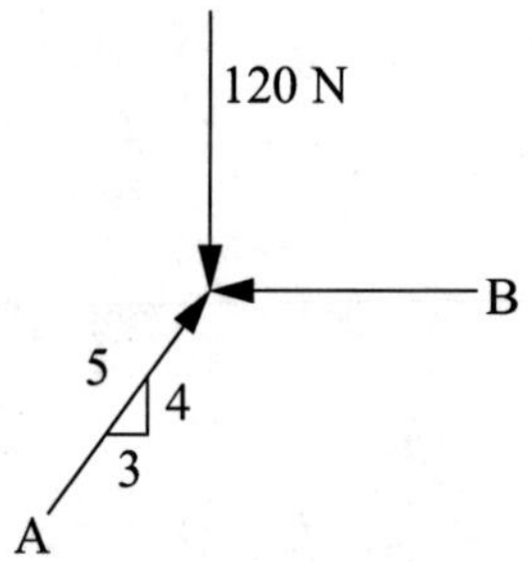

THE CORRECT ANSWER IS: (C)

34. $R_y = \Sigma F_y = \frac{12}{13}(260) + \frac{3}{5}(300) - 50 = 370$

$R_x = \Sigma F_x = -\frac{5}{13}(260) + \frac{4}{5}(300) = 140$

$R = \sqrt{R_x^2 + R_y^2} = \sqrt{370^2 + 140^2}$

$R = 396 \text{ N}$

THE CORRECT ANSWER IS: (D)

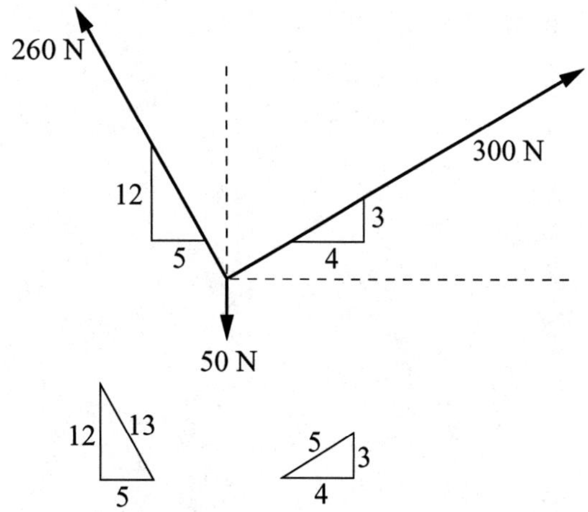

35.

BLOCK A

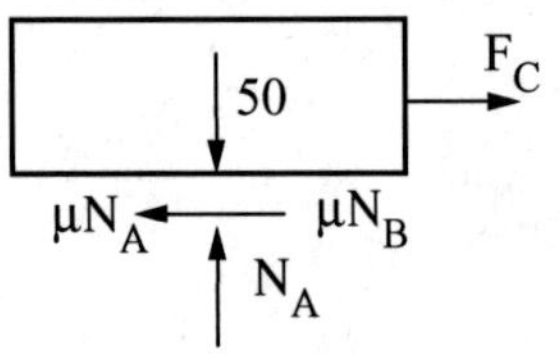

$$\Sigma F_y = 0 = -50 + N_A$$
$$N_A = 50 \text{ N}$$

BLOCK B

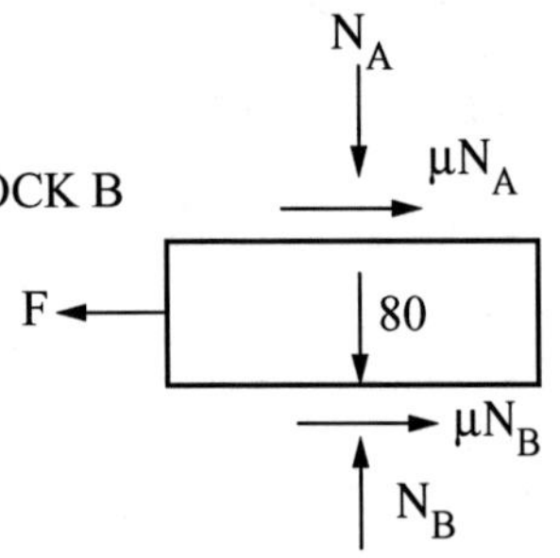

$$\Sigma F_y = 0 = -50 - 80 + N_B$$
$$N_B = 130 \text{ N}$$

BLOCK C

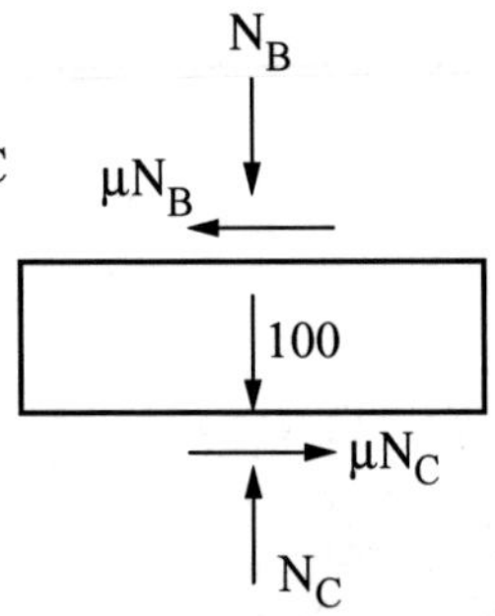

$$\Sigma F_y = 0 = -130 - 100 + N_C$$
$$N_C = 230 \text{ N}$$

Assume Blocks A and C remain stationary.

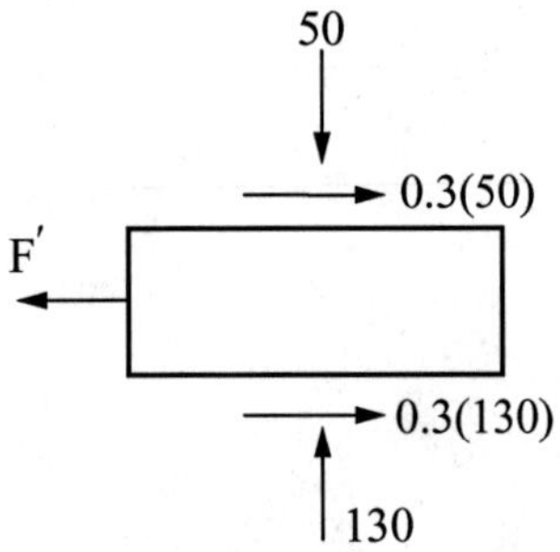

$$\Sigma F_x = 0 = -F' + 0.3(50) + 0.3(130)$$
$$F' = 54 \text{ N}$$

Assume Blocks B and C move.

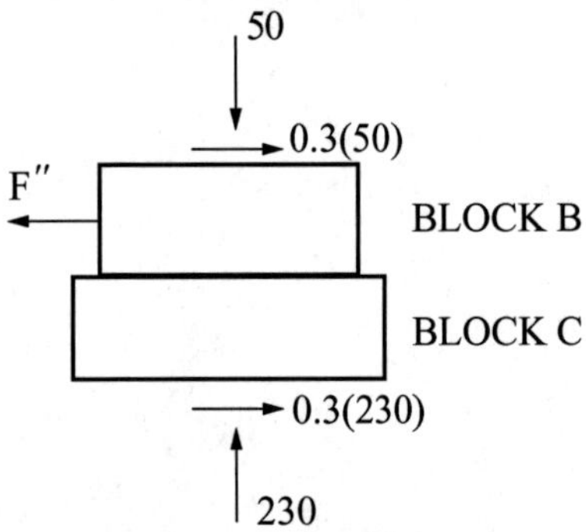

$$\Sigma F_x = 0 = -F'' + 0.3(50) + 0.3(230)$$
$$F'' = 84 \text{ N}$$

$\therefore$ **F** = 54 N where A and C remain stationary.

THE CORRECT ANSWER IS: (B)

36. $F_H = 500 \cos 30° = 433$

$F_V = 500 \sin 30° = 250$

$M_P = 250(0.30) - 433(0.10) = 31.7$ N•m ccw

THE CORRECT ANSWER IS: (A)

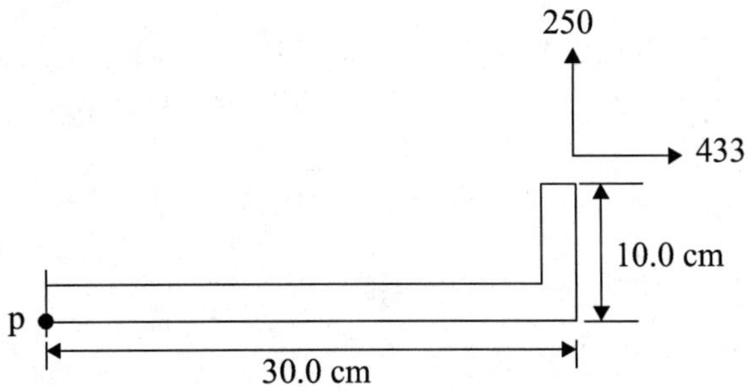

37. Zero-force members usually occur at joints where members are aligned as follows:

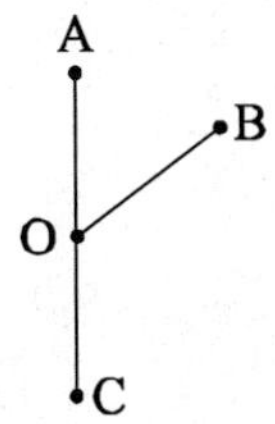

That is, joints where two members are along the same line (OA and OC) and the third member is at some arbitrary angle. That member (OB) is a zero-force member because the forces in OA and OC must be equal and opposite.

For this specific problem, we immediately examine Joints B and E:

BG is zero-force member

CE is zero-force member

Now, examine Joint G. Since BG is zero-force member, the joint effectively looks like:

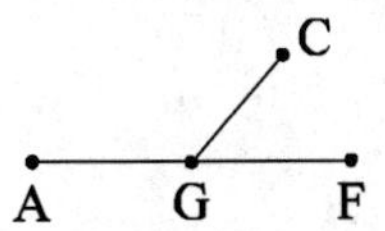

and, therefore, CG is another zero-force member.

Finally, examine Joint C. Since both CG and CE are zero force members, the joint effectively looks like:

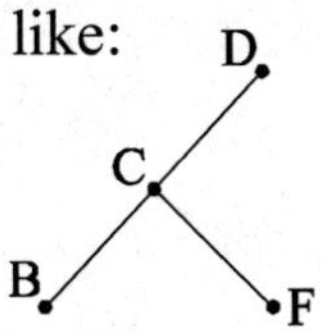

and, therefore, CF is another zero-force member. Thus, BG, CE, CG, CF are the zero-force members.

THE CORRECT ANSWER IS: (A)

38. By definition of a cantilever beam, it is NOT statically indeterminate, it is completely supported, and it is loaded only at a specific point.

THE CORRECT ANSWER IS: (A)

39. $\frac{10\text{ m}}{10\text{ kN}} = \frac{x}{6\text{ kN}}$

$x = 6\text{ m}$

Area 1 = 13(2) = 26 kN•m

Area 2 = $\frac{6(6)}{2}$ = 18 kN•m

Area 3 = 4(4) = 16 kN•m

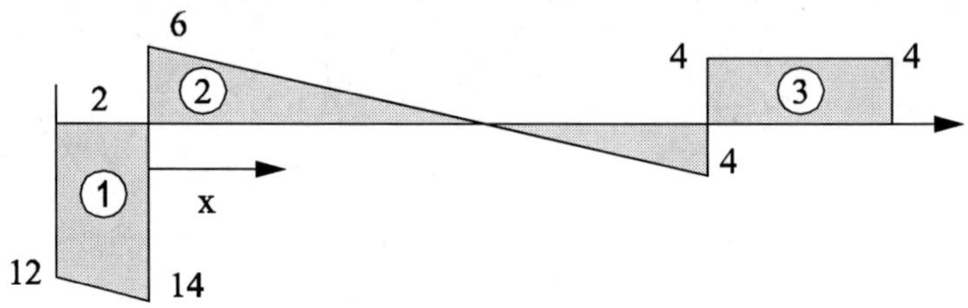

SHEAR DIAGRAM

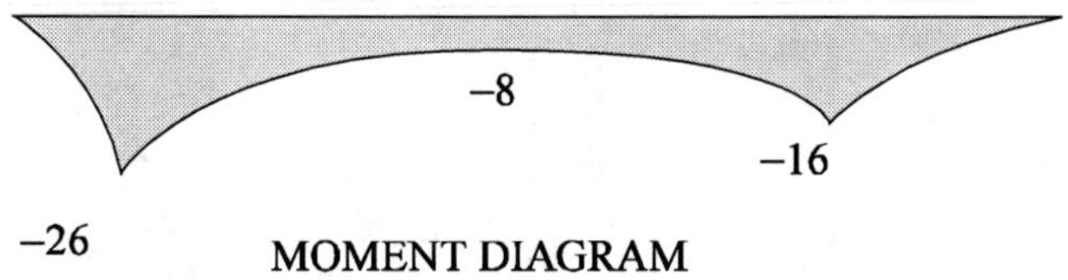

MOMENT DIAGRAM

Maximum magnitude of the bending moment is 26 kN·m.

THE CORRECT ANSWER IS: (D)

40. $\Sigma F = PA = \left(1.4 \times 10^6\right)\left(\frac{\pi(0.5)^2}{4}\right) = F_{rod}$

$F_{rod} = 275\text{ kN} = \sigma A = 68 \times 10^6 A$

$A = 40.4 \times 10^{-4}\text{ m}^2$

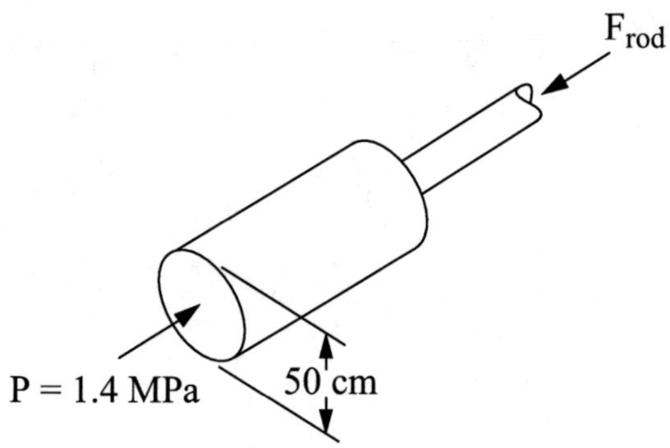

THE CORRECT ANSWER IS: (A)

41. $$\tau = \frac{Tr}{J} = \frac{T\frac{d}{2}}{\frac{\pi d^4}{32}} = \frac{16T}{\pi d^3}$$

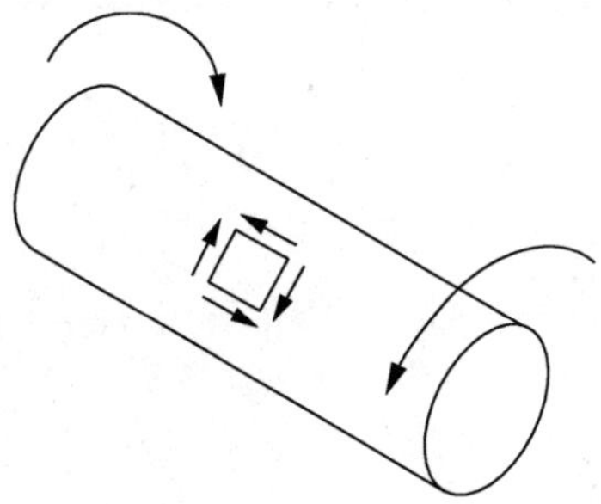

$$T = \frac{\pi d^3 \tau}{16} = \frac{\pi (0.2)^3 (840 \times 10^3)}{16}$$

$$T = 1{,}319 \text{ N}\bullet\text{m}$$

THE CORRECT ANSWER IS: (C)

42. The Euler formula is used for elastic stability of relatively long columns, subjected to concentric axial loads in compression.

THE CORRECT ANSWER IS: (D)

43. The question statement is the definition of hot working.

THE CORRECT ANSWER IS: (A)

44. By definition, a metal with high hardness has a high tensile strength and a high yield strength.

THE CORRECT ANSWER IS: (D)

45. By definition, amorphous materials do not have a crystal structure.

THE CORRECT ANSWER IS: (D)

46. Aluminum is anodic relative to copper and, therefore, will corrode to protect the copper.

THE CORRECT ANSWER IS: (B)

47. The mean pressure of the fluid acting on the gate is evaluated at the mean height, and the center of pressure is 2/3 of the height from the top; thus, the total force of the fluid is:

$$F_f = \rho g \frac{H}{2}(H) = 1,600(9.807)\frac{3}{2}(3) = 70,610 \text{ N}$$

and its point of application is 1.00 m above the hinge. A moment balance about the hinge gives:

$$F(3) - F_f(1) = 0$$

$$F = \frac{F_f}{3} = \frac{70,610}{3} = 23,537 \text{ N}$$

THE CORRECT ANSWER IS: (C)

48. Refer to the Fluid Mechanics section of the *FE Reference Handbook.*

$$\tau_t = \mu\left(\frac{dv}{dy}\right)$$

where τ_t = shear stress and

$\frac{dv}{dy}$ = rate of shear deformation

Hence, μ is the ratio of shear stress to the rate of shear deformation.

THE CORRECT ANSWER IS: (C)

49. $Q = A_1V_1 = (0.01\ m^2)(30\ m/s)$

$= 0.3\, m^3/s$

Since the water jet is deflected perpendicularly, the force F must deflect the total horizontal momentum of the water.

$F = \rho QV = (1{,}000\ kg/m^3)\ (0.3\ m^3/s)\ (30\ m/s) = 9{,}000\ N = 9.0\ kN$

THE CORRECT ANSWER IS: (B)

50. Flow through an insulated valve closely follows a throttling process. A throttling process is at constant enthalpy.

THE CORRECT ANSWER IS: (B)

51. $\frac{\rho v^2}{2} = gh(\rho - \rho_{air})$

$$\therefore h = \frac{\rho v^2}{2g(\rho - \rho_{air})} \approx \frac{v^2}{2g} \approx \frac{(2)^2}{(2)(9.8)} \approx 0.204\ m$$

THE CORRECT ANSWER IS: (D)

52. S = apparent power
P = real power
Q = reactive power

$S = P + jQ = |S| \cos\theta + j\,|S| \sin\theta$

$\cos\theta = pf = 0.866$

$Q = (1{,}500\ VA)\ \sin[\cos^{-1} 0.866] = 750\ VAR$

THE CORRECT ANSWER IS: (B)

53. $Z = 30 + j90 - j50 = 30 + j40\ \Omega$

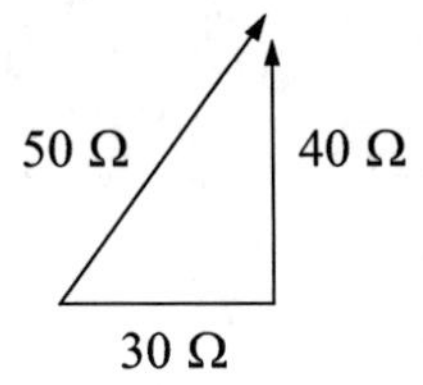

THE CORRECT ANSWER IS: (D)

54. Initially, $V_C(t) = 150$ V

$$W_C(t) = \frac{1}{2}cV_C^2 = \frac{1}{2}(10 \times 10^{-6} F)(150\text{ V})^2$$

$W_C = 0.113$ J initial stored energy.

After ten time constants, all energy will be dissipated.

THE CORRECT ANSWER IS: (B)

55. $R_T = 4\Omega + 3\Omega \,||\, 6\Omega = 4\Omega + 2\Omega$

$$R_T = 6\Omega \Rightarrow I_T = \frac{6\text{ V}}{6\Omega} = 1\text{ A}$$

$$I_x = \frac{3}{9}(I_T) = \frac{1}{3}\text{A}$$

THE CORRECT ANSWER IS: (A)

56. $$\frac{(1-i)^2}{(1+i)^2} = \frac{1 - 2i + i^2}{1 + 2i + i^2} = \frac{1 - 1 - 2i}{1 - 1 + 2i} = \frac{-i}{i} = -1$$

THE CORRECT ANSWER IS: (A)

57. As vapor escapes, the mass within the tank is reduced. With constant volume, the specific volume within the tank must increase. This can happen only if liquid evaporates.

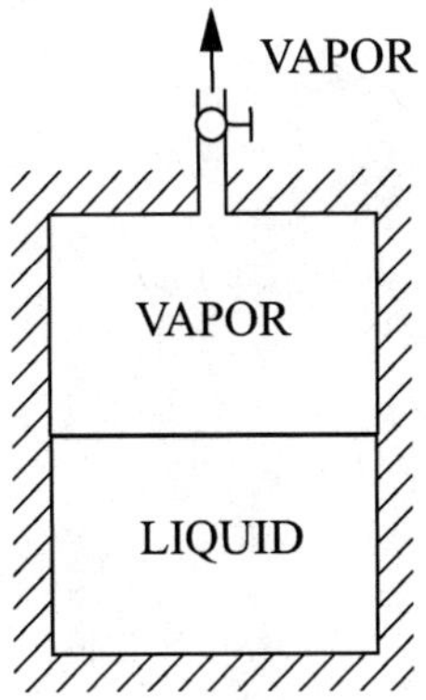

THE CORRECT ANSWER IS: (A)

58. By definition, an adiabatic process is a process in which no heat is transferred.

THE CORRECT ANSWER IS: (D)

59. An isentropic process is one for which the entropy remains constant. Entropy is defined by the equation:

$$dS = \left(\frac{\delta Q}{T}\right)_{\text{reversible}}$$

The entropy will be constant if $\delta Q = 0$ and the process is reversible. It is theoretically possible for a nonadiabatic, irreversible process to have a constant entropy, but this is not one of the responses. Option (D) describes a state, not a process.

THE CORRECT ANSWER IS: (C)

60. $$R = \frac{Ru}{M} = \frac{8.314}{44} = 0.1890 \frac{kJ}{kg \bullet K}$$

THE CORRECT ANSWER IS: (A)

EXAM SPECIFICATIONS FOR THE AFTERNOON SESSION

AFTERNOON SESSION IN ELECTRICAL ENGINEERING: (60 questions in 9 topic areas)

Topic Area	Approximate Percentage of Test Content
I. Circuits	**16%**
A. KCL, KVL	
B. Series/parallel equivalent circuits	
C. Node and loop analysis	
D. Thevenin/Norton theorems	
E. Impedance	
F. Transfer functions	
G. Frequency/transient response	
H. Resonance	
I. Laplace transforms	
J. 2-port theory	
K. Filters (simple passive)	
II. Power	**13%**
A. 3-phase	
B. Transmission lines	
C. Voltage regulation	
D. Delta and wye	
E. Phasors	
F. Motors	
G. Power electronics	
H. Power factor (pf)	
I. Transformers	
III. Electromagnetics	**7%**
A. Electrostatics/magnetostatics (e.g., measurement of spatial relationships, vector analysis)	
B. Wave propagation	
C. Transmission lines (high frequency)	
IV. Control Systems	**10%**
A. Block diagrams (feed forward, feedback)	
B. Bode plots	
C. Controller performance (gain, PID), steady-state errors	
D. Root locus	
E. Stability	

Topic Area		Approximate Percentage of Test Content
V.	**Communications**	**9%**
	A. Basic modulation/demodulation concepts (e.g., AM, FM, PCM)	
	B. Fourier transforms/Fourier series	
	C. Sampling theorem	
	D. Computer networks, including OSI model	
	E. Multiplexing	
VI.	**Signal Processing**	**8%**
	A. Analog/digital conversion	
	B. Convolution (continuous and discrete)	
	C. Difference equations	
	D. Z-transforms	
VII.	**Electronics**	**15%**
	A. Solid-state fundamentals (tunneling, diffusion/drift current, energy bands, doping bands, p-n theory)	
	B. Bias circuits	
	C. Differential amplifiers	
	D. Discrete devices (diodes, transistors, BJT, CMOS) and models and their performance	
	E. Operational amplifiers	
	F. Filters (active)	
	G. Instrumentation (measurements, data acquisition, transducers)	
VIII.	**Digital Systems**	**12%**
	A. Numbering systems	
	B. Data path/control system design	
	C. Boolean logic	
	D. Counters	
	E. Flip-flops	
	F. Programmable logic devices and gate arrays	
	G. Logic gates and circuits	
	H. Logic minimization (SOP, POS, Karnaugh maps)	
	I. State tables/diagrams	
	J. Timing diagrams	
IX.	**Computer Systems**	**10%**
	A. Architecture (e.g., pipelining, cache memory)	
	B. Interfacing	
	C. Microprocessors	
	D. Memory technology and systems	
	E. Software design methods (structured, top-down bottom-up, object-oriented design)	
	F. Software implementation (structured programming, algorithms, data structures)	

ELECTRICAL
AFTERNOON SAMPLE QUESTIONS

NOTE: THESE QUESTIONS REPRESENT ONLY HALF THE NUMBER OF QUESTIONS THAT APPEAR ON THE ACTUAL EXAMINATION.

ELECTRICAL SAMPLE QUESTIONS

1. Consider the following circuit:

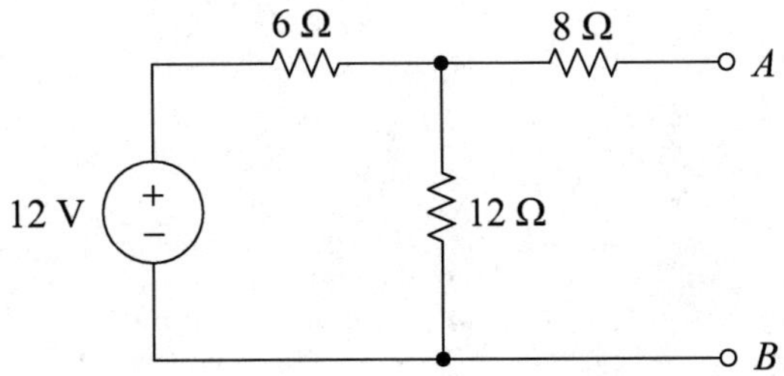

The Thevenin equivalent resistance (Ω) at points *A*–*B* is most nearly:

(A) 8
(B) 12
(C) 20
(D) 26

GO ON TO THE NEXT PAGE

2. Consider the following network:

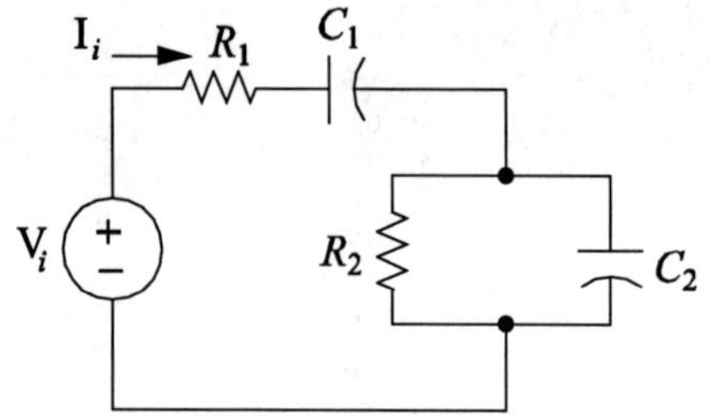

The driving point impedance (input impedance) has poles at:

(A) $s = 0$ and $s = -\frac{1}{R_2C_2}$

(B) $s = -\frac{1}{R_1C_1}$ and $s = -\frac{1}{R_2C_2}$

(C) $s = 0$ and $s = -\frac{1}{R_1C_2}$

(D) $s = -\frac{1}{R_1C_2}$ and $s = -\frac{1}{R_2C_1}$

3. Refer to the figure below.

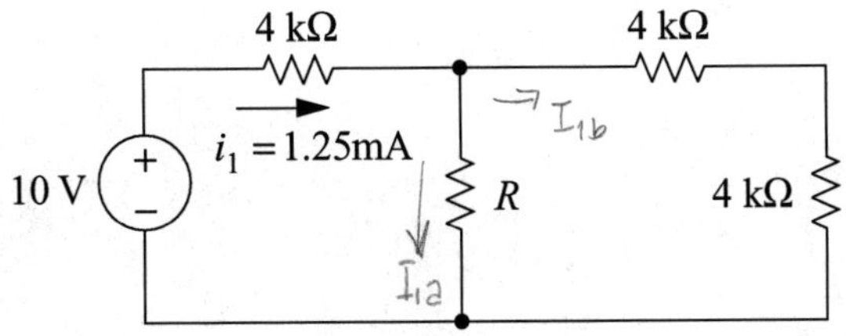

The value of R (kΩ) needed to make i_1 equal 1.25 mA is most nearly:

(A) 0
(B) 4
(C) 8
(D) 12

GO ON TO THE NEXT PAGE

Questions 4–5 refer to the circuit shown in the figure below. After having been closed for a long time, the switch is opened at $t = 0$.

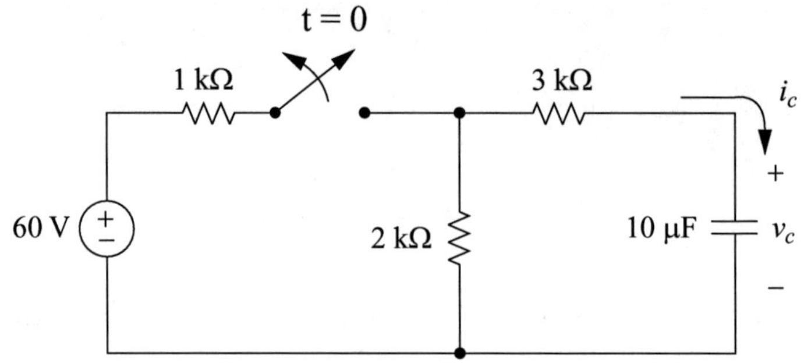

4. The expression for v_c (V) for $t > 0$ is most nearly:

(A) $20\ e^{-50t}$
(B) $20\ e^{-0.2t}$
(C) $40\ e^{-200{,}000t}$
(D) $40\ e^{-20t}$

5. The energy (mJ) dissipated in the 2-kΩ resistor for the period $t = 0^+$ to $t = \infty$ is most nearly:

(A) 1.6
(B) 3.2
(C) 6.0
(D) 8.0

GO ON TO THE NEXT PAGE

Questions 6–7: A balanced 3-phase source can be represented by three wye-connected generators, as shown below.

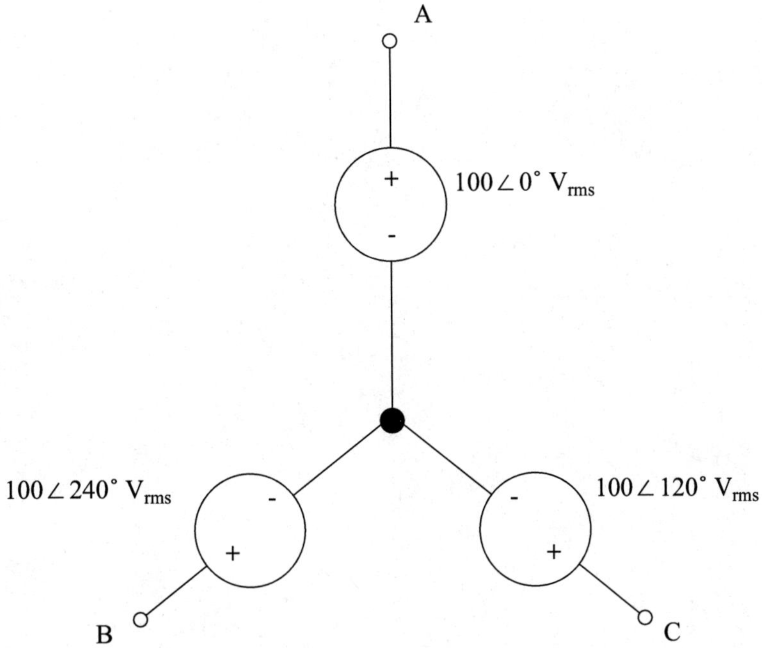

6. The line-to-line voltage V_{AB} (V_{rms}) is most nearly:

(A) 173.2∠ 30°

(B) 173.2∠–30°

(C) 57.7∠ 30°

(D) 57.7∠–30°

7. The source is connected by three lines to a 3-phase delta-connected load with 10 Ω per phase. The rms current (amperes) in each of the lines is most nearly:

(A) 10.0
(B) 17.3
(C) 30.0
(D) 52.0

GO ON TO THE NEXT PAGE

8. A pump station uses an induction motor that requires a complex power of

$\mathbf{S}_1 = 20 \angle 36.87° \text{ kVA}$

and a synchronous motor that requires a complex power of

$\mathbf{S}_2 = 10 \angle -53.13° \text{ kVA}$.

The reactive power (kVAR) required by the pump station is most nearly:

(A) 4
(B) 20
(C) 22
(D) 30

9. The armature circuit of a dc motor may be modeled as a voltage source in series with a resistor. The field of the motor is supplied by a shunt field winding as shown below.

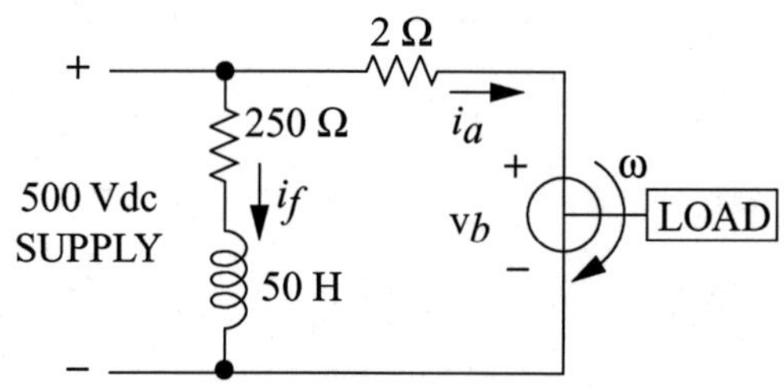

For a particular field strength, the back emf v_b is directly proportional to the motor speed ω. In steady state, under a particular load, the current i_a is measured to be 12 A when ω is 850 rpm and i_f is 2 A.

With field current at 2 A, if the motor were started with no additional resistance in series with the armature, the initial current i_a (amperes) would be most nearly:

(A) –12
(B) 0
(C) 12
(D) 250

10. The following circuit contains a lossless transmission line with a characteristic impedance of $Z_O = 50\ \Omega$. The magnitude of the load voltage V_L is 200 V_{rms}.

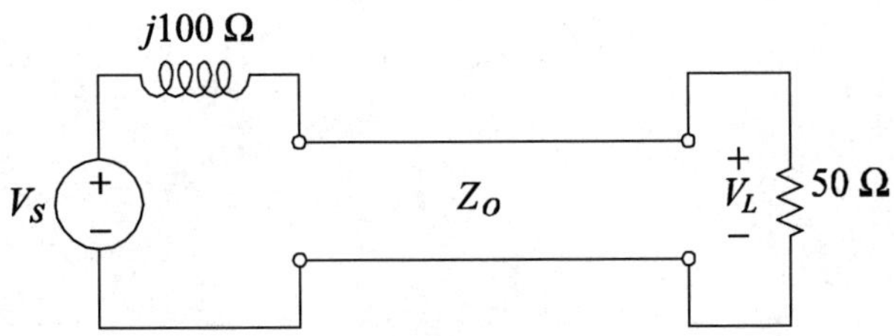

The magnitude of the source voltage V_s (V_{rms}) is most nearly

(A) 100.0
(B) 200.0
(C) 282.8
(D) 447.2

11. The solid cylindrical conductor shown below carries a uniform direct current having a density of 100 A/m^2 in the positive z direction. Assume the resistivity is 0.1 Ω·m.

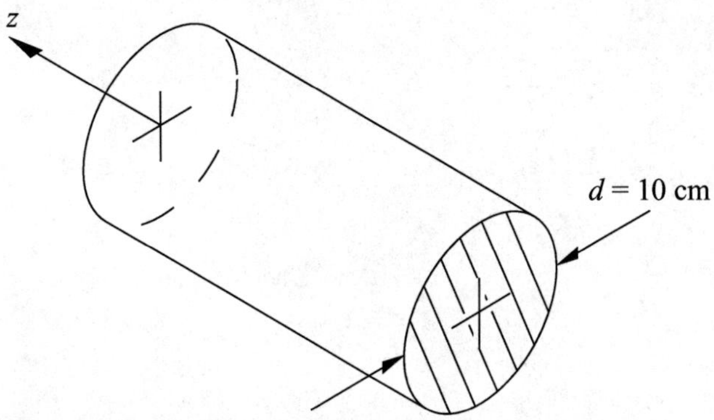

NOT TO SCALE

The power loss (W) per meter length is most nearly:

(A) 2.50
(B) 3.93
(C) 7.85
(D) 31.42

Questions 12–13: A proportional controller with gain K is used to control a spring and mass system as shown in the figure below.

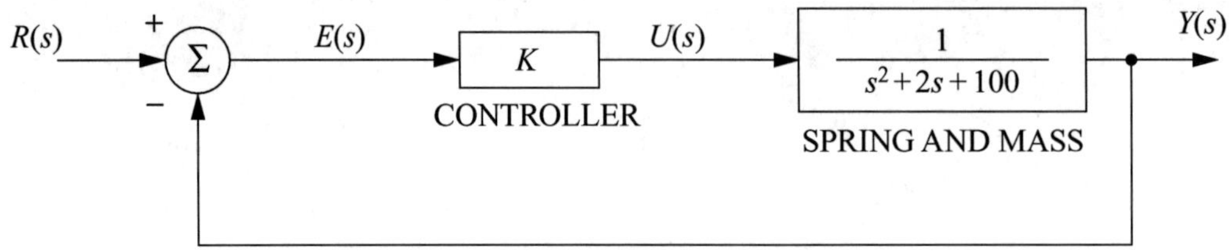

12. The value of K that will result in the closed-loop system characteristic polynomial $P(s) = s^2 + 2s + 150$ is most nearly:

(A) 0
(B) 25
(C) 50
(D) 100

13. If K is adjusted so that the closed-loop characteristic polynomial is $s^2 + 2s + 150$, then the system damping ratio is most nearly:

(A) 0.8
(B) 0.08
(C) 0.0067
(D) 0

GO ON TO THE NEXT PAGE

14. A unity-feedback control system is shown in the figure below.

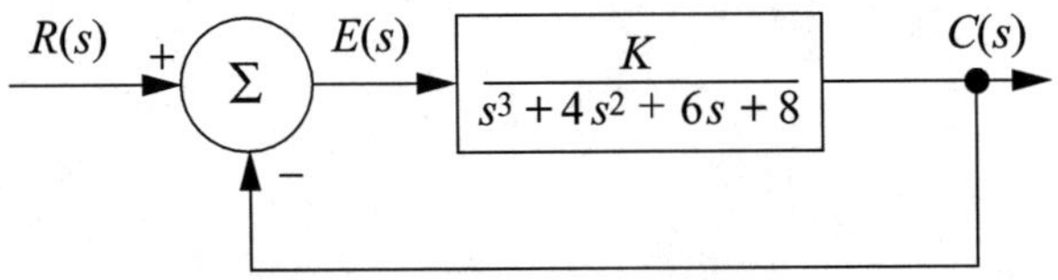

The range of K for which this system is stable is most nearly:

(A) $0 < K < 16$
(B) $8 < K < 16$
(C) $-8 < K < 16$
(D) $-16 < K < -8$

15. A 10-MHz carrier is AM-modulated by a symmetrical square wave with a period of 1 ms. The bandwidth (kHz) of a filter with a center frequency of 10 MHz required to transmit the square wave with the first five nonzero components ($n = 1, 3, 5, 7, 9$) of its Fourier components is most nearly:

(A) 3
(B) 6
(C) 10
(D) 20

GO ON TO THE NEXT PAGE

16. An amplitude modulation system is shown below in block diagram form.

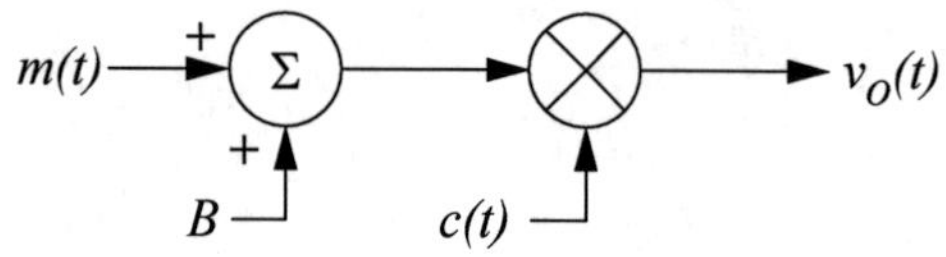

$m(t) = A_m \cos \omega_m t$
B is a constant bias voltage.
$c(t) = A_c \cos \omega_c t$

The frequencies found in the output $v_o(t)$ are:

(A) ω_c, ω_m

(B) ω_c, $\omega_c + \omega_m$

(C) $\omega_c \pm \omega_m$

(D) ω_c, $\omega_c \pm \omega_m$

17. Two discrete time signals ($x[n]$ and $y[n]$) are defined as shown in the table with $x[n] = y[n] = 0$ for all values of n less than –1 and for all values of n greater than 3.

n	$x[n]$	$y[n]$
–1	0	–1
0	1	2
1	2	1
2	–1	1
3	–2	0

The discrete convolution $v[n] = x[n] * y[n]$ for $n = –2$ to $+5$ is most nearly:

(A) [0, –1, 1, 2, 2, 1, –1, 1]

(B) [0, 0, –1, 2, 2, –1, 0, 0]

(C) [–6, 0, 3, 3, 1, 0, 0, 0]

(D) [0, –1, 0, 6, 3, –3, –3, –2]

18. The function $x(t)$ is to be convolved with $h(t)$. Both functions are shown below.

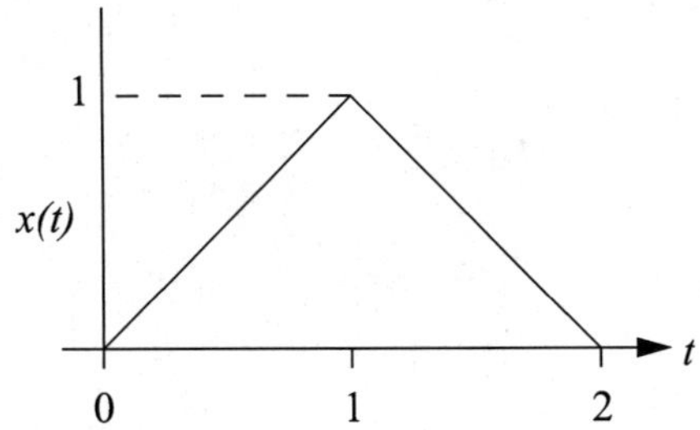

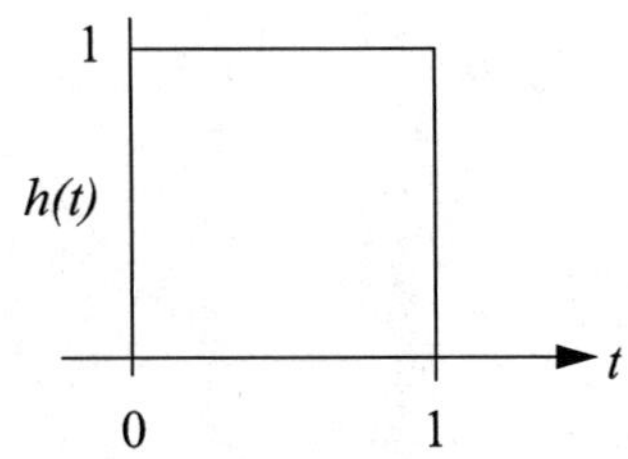

The value of the convolution of $x(t)$ with $h(t)$ for $0 < t < 1$ is most nearly:

(A) $t/2$
(B) t
(C) $t^2/2$
(D) t^2

19. A digital filter with input $x[k]$ and output $y[k]$ is described by the difference equation

$$y[k] = \frac{1}{6}\left(3x[k] + 2x[k-1] + x[k-2]\right)$$

The discrete-time transfer function of the filter $H(z)$ is:

(A) $\dfrac{6z^2}{3z^2 + 2z + 1}$

(B) $\dfrac{6z}{3z^2 + 2z + 1}$

(C) $3z^2 + 2z + 1$

(D) $\dfrac{1}{6}\left[\dfrac{3z^2 + 2z + 1}{z^2}\right]$

20. An ideal operational amplifier is to be connected as a differential amplifier shown in the diagram below. Nominally, $R_1 = R_3 = 5\ \text{k}\Omega$ and $R_2 = R_4 = 200\ \text{k}\Omega$.

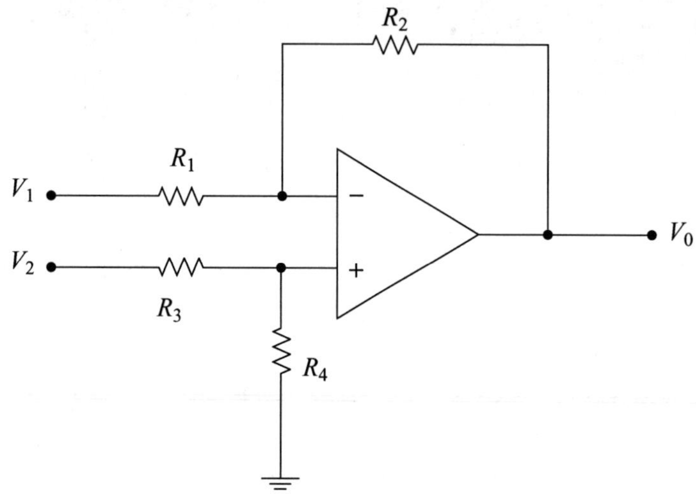

The gain for an input V_2 with the input V_1 grounded is most nearly:

(A) 20
(B) 35
(C) 40
(D) 41

GO ON TO THE NEXT PAGE

21. The figure below shows a circuit with an *n*-channel enhancement mode MOSFET transistor.

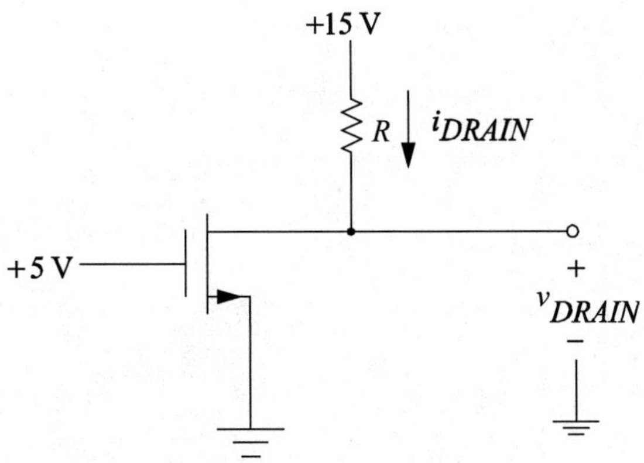

In the FET triode region:

$$i_{DRAIN} = K[2(v_{GS} - V_t)v_{DS} - v_{DS}^2]$$

In the FET saturation region:

$$i_{DRAIN} = K(v_{GS} - V_t)^2$$

The FET operates in saturation for $v_{DS} \geq v_{GS} - V_t$.

Assume that $K = 0.5\ \text{mA/V}^2$ and $V_t = 1$ V.

If $v_{DRAIN} = 2$ V, the value of R (kΩ) is most nearly:

(A) 1.63
(B) 2.17
(C) 2.50
(D) 3.42

GO ON TO THE NEXT PAGE

22. The transistor in the circuit below has a very high value of β.

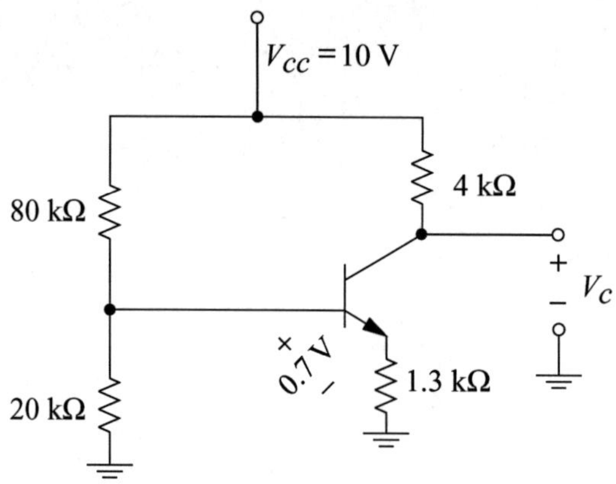

The value of the collector voltage V_c (V) is most nearly:

(A) 6.0
(B) 4.0
(C) 2.0
(D) 1.3

23. For a *pn* junction, the contact potential V_0 is:

(A) a built-in potential barrier that is necessary for equilibrium at the junction

(B) a measure of the ohmic resistance of the junction

(C) independent of charge carrier concentrations in the p and n regions

(D) independent of temperature

24. The periodic waveform with period T shown below is the input to the circuit shown with an ideal op amp.

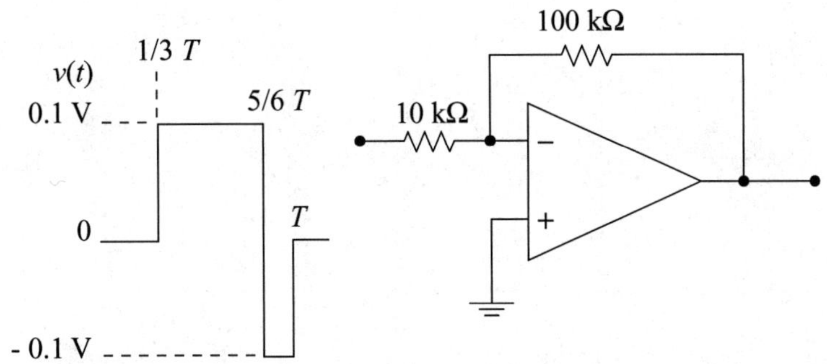

The dc component of the output of the circuit (V) in steady-state operation is most nearly:

(A) 0.333
(B) 0
(C) –0.333
(D) –1.414

25. Consider the following Karnaugh map:

AB \ CD	00	01	11	10
00	1			1
01	1	1	1	1
11				
10	1			1

Which logic function best represents a minimal SOP expression?

(A) $f(A,B,C,D) = \overline{B}\,\overline{D} + \overline{A}\,B + \overline{A}\,\overline{D}$

(B) $f(A,B,C,D) = \overline{A}\,\overline{D} + \overline{A}\,B + A\,\overline{B}\,\overline{D}$

(C) $f(A,B,C,D) = \overline{A}\,\overline{D} + \overline{A}\,B\,D + A\,\overline{B}\,\overline{D}$

(D) $f(A,B,C,D) = \overline{B}\,\overline{D} + \overline{A}\,B$

GO ON TO THE NEXT PAGE

26. Flip-flops A and B form a sequential synchronous circuit as shown below.

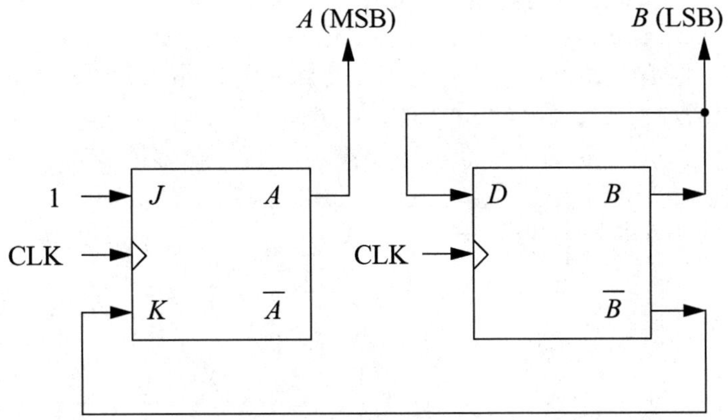

After the clock pulse, binary count 10 ($A = 1$, $B = 0$) changes to:

(A) 00
(B) 01
(C) 10
(D) 11

27. Which of the following is a binary representation of the base-10 fraction $\frac{93}{128}$?

(A) 0.1011100
(B) 0.1011101
(C) 0.1011110
(D) 0.1011111

GO ON TO THE NEXT PAGE

28. When a CPU fetches an instruction word from memory, the word contains an operation code (op code) that indicates the type of operation the CPU is to perform and information specifying where the instruction operand(s) is (are) located. A computer may use various addressing modes to specify the operand location. One such addressing mode is illustrated below, where R designates some register within the CPU and d is a constant embedded in the instruction word.

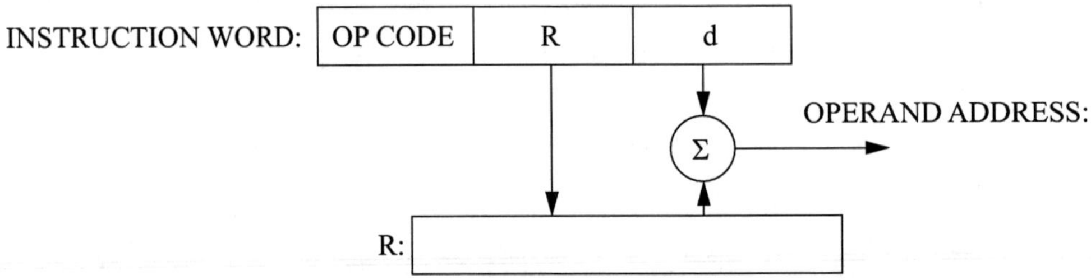

Which of the following terms best describes the addressing mode used by the instruction above?

(A) Immediate addressing

(B) Direct addressing

(C) Indexed addressing

(D) Indirect addressing

GO ON TO THE NEXT PAGE

29. A microprocessor (μp) uses a 16-bit address bus and an 8-bit data bus. The address bus lines are labeled A_{15} to A_0, where A_{15} is the most significant address bit and A_0 is the least significant address bit. The microprocessor generates an active-low address strobe (AS) at the beginning of each memory access cycle to indicate that a valid address has been placed on the address bus. The AS signal is asserted for the entire memory access cycle.

In a certain application of the μp, the three most significant address bits (A_{15}–A_{13}) are used as inputs to a 3-of-8 decoder to generate chip-select signals (CS_0–CS_7). The chip-select signals are used to enable an EPROM memory device whose address lines are tied directly to the lower 13 address lines (A_{12}–A_0) of the μp. The EPROM's output lines are enabled whenever its enable input (EN) is pulled low. A figure of the microprocessor described above and the decoder truth table are shown below.

The range of addresses (expressed in hexadecimal) to which the EPROM in the figure will respond is most nearly:

(A) 0000 to 1FFF
(B) 4000 to 5FFF
(C) 8000 to 9FFF
(D) E000 to FFFF

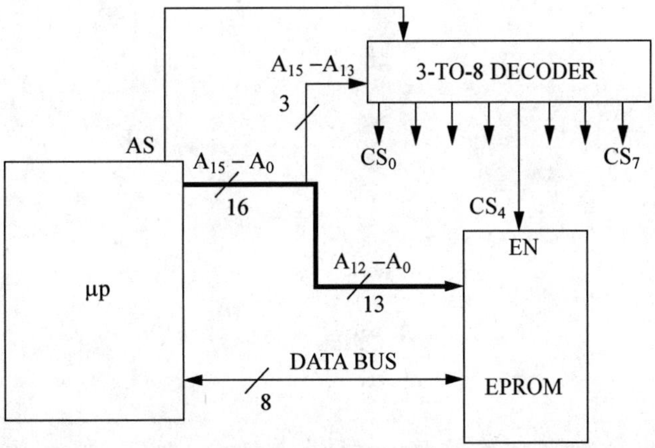

Decoder Truth Table

AS	A_{15}	A_{14}	A_{13}	CS_0	CS_1	CS_2	CS_3	CS_4	CS_5	CS_6	CS_7
1	X	X	X	1	1	1	1	1	1	1	1
0	0	0	0	0	1	1	1	1	1	1	1
0	0	0	1	1	0	1	1	1	1	1	1
0	0	1	0	1	1	0	1	1	1	1	1
0	0	1	1	1	1	1	0	1	1	1	1
0	1	0	0	1	1	1	1	0	1	1	1
0	1	0	1	1	1	1	1	1	0	1	1
0	1	1	0	1	1	1	1	1	1	0	1
0	1	1	1	1	1	1	1	1	1	1	0

X = "don't care"

30. An analog-to-digital converter (ADC) is used to collect 512 samples of a voltage signal. The samples are taken at intervals T of 1 μs and stored in an array S where S(0) holds the first sample point, S(1) the second sample point, and so on. The following flow diagram describes the logic of a computer program used to process the data.

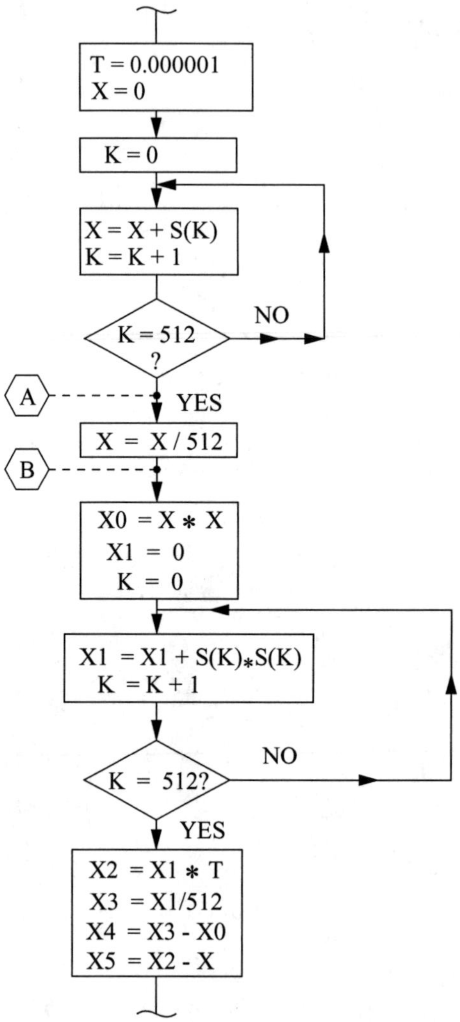

The AC power of the sample signal is defined by the following expression where N is the number of samples.

$$P_S = \frac{1}{N}\sum_{k=0}^{N-1} S^2(k) - \left[\frac{1}{N}\sum_{k=0}^{N-1} S(k)\right]^2$$

By the end of the program, which variable represents the value of P_S?

(A) X0
(B) X3
(C) X4
(D) X5

IF YOU FINISH BEFORE TIME IS CALLED, YOU MAY CHECK YOUR WORK ON THIS TEST.

ELECTRICAL AFTERNOON SOLUTIONS

ANSWERS TO THE ELECTRICAL AFTERNOON SAMPLE QUESTIONS

Detailed solutions for each question begin on the next page.

QUESTION	ANSWER
1	B
2	A
3	C
4	D
5	B
6	A
7	C
8	A
9	D
10	D
11	C
12	C
13	B
14	C
15	D
16	D
17	D
18	C
19	D
20	C
21	B
22	A
23	A
24	C
25	D
26	A
27	B
28	C
29	C
30	C

ELECTRICAL AFTERNOON SOLUTIONS

1. With the 12 V source zeroed (short), $R_{TH} = 8 + 12 \| 6 = 8 + 4 = 12\ \Omega$

THE CORRECT ANSWER IS: (B)

2.

$$\mathbf{Z}_{in} = \frac{V_i}{I_i} = R_1 + \frac{1}{sC_1} + \frac{\frac{R_2}{sC_2}}{R_2 + \frac{1}{sC_2}}$$

$$= \frac{sR_1C_1 + 1}{sC_1} + \frac{R_2}{sR_2C_2 + 1}$$

$$= \frac{(sR_1C_1 + 1)(sR_2C_2 + 1) + sR_2C_1}{R_2C_1C_2\left[s\left(s + \frac{1}{R_2C_2}\right)\right]}$$

Poles at $s = 0$; $s = -\frac{1}{R_2C_2}$

THE CORRECT ANSWER IS: (A)

3. If $i_1 = 1.25$ mA, then the total resistance i_1 can see must be $\frac{10\ \text{V}}{1.25\ \text{mA}} = 8\ \text{k}\Omega$

$$4{,}000 + \frac{R(4{,}000 + 4{,}000)}{R + 4{,}000 + 4{,}000} = 8{,}000\ \Omega$$

$$\frac{8{,}000R}{8{,}000 + R} = 4{,}000\ \Omega$$

$$8{,}000R = 4{,}000R + 32\ \text{M}\Omega$$

$$R = 32\ \text{M}\Omega / 4\ \text{k}\Omega$$

$$= 8\ \text{k}\Omega$$

THE CORRECT ANSWER IS: (C)

4. After a long time, $v_c = 60 \times \frac{2}{2+1} = 40\text{ V}$

After $t = 0^+$,

$$v_c(t) = v_c(0)\,e^{\frac{-t}{RC}}$$

$$RC = 5\text{ k}\Omega \times 10\ \mu\text{F} = 5{,}000 \times 10^{-5} = 0.05$$

$$v_c(t) = 40\,e^{\frac{-t}{0.05}} = 40\,e^{-20t}$$

THE CORRECT ANSWER IS: (D)

5. $W_c = 1/2\ C\ V_c^2 = 1/2\ (10\ \mu\text{F})(40\text{ V})^2$

$W_c = 5 \times 10^{-6}\ (1{,}600)$

$W_c = 8\text{ mJ}$

The 2-kΩ resistor dissipates $\frac{2}{2+3}$ = 40% of W_c or 3.2 mJ

THE CORRECT ANSWER IS: (B)

6. $V_{AB} = V_{AN} - V_{BN} = 100\angle 0° - 100\angle 240° = 173.2\angle 30°$

THE CORRECT ANSWER IS: (A)

7. At the load,

$$I_\phi = \frac{\sqrt{3} \times 100\text{ V}}{10\Omega} = 17.32\text{ A}$$

$$I_1 = \sqrt{3} \times 17.32 = 30\text{ A}$$

THE CORRECT ANSWER IS: (C)

8. $\mathbf{S}_1 = 20\ \angle 36.87° = 16 + j12$ kVA

$\mathbf{S}_2 = 10\ \angle{-53.13°} = 6 - j8$ kVA

$\mathbf{S} = \mathbf{S}_1 + \mathbf{S}_2 = (16 + j12) + (6 - j8)$

$\mathbf{S} = 22 + j4 = P + jQ \Rightarrow Q = 4$ kVAR

THE CORRECT ANSWER IS: (A)

9. $$i_a = \frac{500\text{ V}}{2\,\Omega} = 250\text{ A}$$

THE CORRECT ANSWER IS: (D)

10.

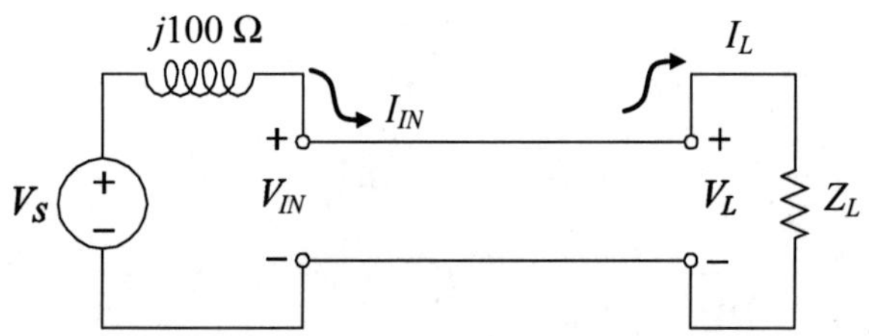

For a matched lossless transmission line ($\mathbf{Z}_L = \mathbf{Z}_0$), $V_{IN} = V_L$ and $I_{IN} = I_L$.

$V_{IN} = 200\ \text{V}_{\text{rms}}$

$I_{IN} = (200/50) = 4\ \text{A}_{\text{rms}}$

$\mathbf{V}_S = 200 + 400j$

$$|\mathbf{V}_S|_{\text{rms}} = \sqrt{200^2 + 400^2} = 447.2\ \text{V}_{\text{rms}}$$

THE CORRECT ANSWER IS: (D)

11. Since the current density is 100 A/m^2, the current in the conductor is given by:

$$I = 100 \times \text{Area} \;=\; 100 \times (\pi r^2) = \;100 \times \pi (0.05)^2$$

$$I = 0.7854\,\text{A}$$

The resistance per unit length is given by:

$$R = \frac{\rho L}{A} = \frac{0.1\,\Omega\text{m} \times \text{L}}{\pi\left(0.05\,\text{m}\right)^2} = 12.73\,\Omega/\text{m}$$

Since $P = I^2 R$, the power loss per unit length is given by:

$$P = (0.785\text{A})^2 \times 12.73\,\Omega/\text{m} = 7.854\,\text{W/m}$$

THE CORRECT ANSWER IS: (C)

12. Refer to Control Systems in the Measurement and Controls section of the *FE Reference Handbook.*

$$\frac{Y(s)}{R(s)} = \frac{\dfrac{K}{s^2+2s+100}}{1+\dfrac{K}{s^2+2s+100}} = \frac{K}{s^2+2s+(K+100)}$$

$$\therefore\quad P(s) = s^2+2s+150 = s^2+2s+(K+100) \;\Rightarrow\; K = 50$$

THE CORRECT ANSWER IS: (C)

13. Refer to Control Systems in the Measurement and Controls section of the *FE Reference Handbook.*

$$P(s) = s^2 + 2\zeta\omega_n s + \omega_n^2$$

$$\therefore\quad \omega_n^2 = 150 \;\Rightarrow\; \omega_n = \sqrt{150} \;\text{ and }\; \zeta = \frac{2}{2\sqrt{150}} = 0.082$$

THE CORRECT ANSWER IS: (B)

14. The closed-loop characteristic equation is:

$$s^3 + 4s^2 + 6s + 8 + K = 0$$

Routh Array

s^3	1	6
s^2	4	$8 + K$
s^1	$\dfrac{24 - 8 - K}{4}$	
s^0	$8 + K$	

From the first column, $-8 < K < 16$

THE CORRECT ANSWER IS: (C)

15.

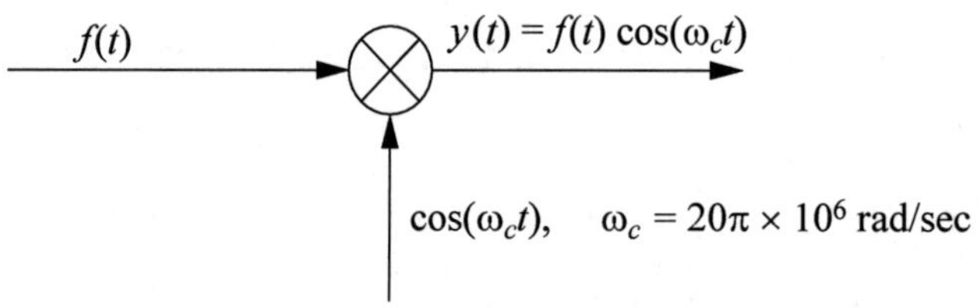

The input signal $f(t)$ is a square wave with period $T = 1$ ms. From the Electrical and Computer Engineering section of the *FE Reference Handbook*, the Fourier series expansion of $f(t)$ is given by:

$$f(t) = \sum_{\substack{n=1 \\ n \ \text{odd}}}^{\infty} (-1)^{(n-1)/2} \frac{4V_0}{n\pi} \cos(n\omega_0 t)$$

where $\omega_0 = \dfrac{2\pi}{T} = \dfrac{2\pi}{1\,\text{ms}} = 2\pi \times 10^3 \quad \text{rad/sec}$

$f_0 = 1$ kHz

The signal $y(t)$ will contain frequency components at $f_c \pm f_0, f_c \pm 3f_0, f_c \pm 5f_0, \ldots$

If the first five non-zero components are to be passed by the filter, the bandwidth of the filter must be $\geq 2 \times 9$ kHz = 18 kHz.

THE CORRECT ANSWER IS: (D)

16. Inputs to multiplier are $B + A_m\cos\omega_m t$ and $A_c\cos\omega_c t$.

Using trigonometric identity:

$v_0(t) = (B + A_m \cos \omega_m t)(A_c \cos \omega_c t)$

$$= BA_c \cos \omega_c t + \frac{A_m A_c}{2} \cos\left(\omega_c + \omega_m\right)t + \frac{A_m A_c}{2} \cos\left(\omega_c - \omega_m\right)t$$

THE CORRECT ANSWER IS: (D)

17. $x[n]{*}y[n] \quad = (\delta[n] + 2\delta[n-1] - \delta[n-2] - 2\delta[n-3]){*}y[n]$

$= y[n] + 2y[n-1] - y[n-2] - y[n-3]$

$$\begin{aligned}= -\delta[n+1] + 2\delta[n] + \delta[n-1] + \delta[n-2] \\ - 2\delta[n] + 4\delta[n-1] + 2\delta[n-2] + 2\delta[n-3] \\ + \delta[n-1] - 2\delta[n-2] - \delta[n-3] - \delta[n-4] \\ + 2\delta[n-2] - 4\delta[n-3] - 2\delta[n-4] - 2\delta[n-5]\end{aligned}$$

$$= -\delta[n+1] + 0\delta[n] + 6\delta[n-1] + 3\delta[n-2] - 3\delta[n-3] - 3\delta[n-4] - 2\delta[n-5]$$

The convolution can be done more abstractly as:

k	–1	0	1	2	3				
y	–1	2	1	1	0				
x	0	1	2	–1	–2				
	0	0	0	0	0				
		–1	2	1	1	0			
			–2	4	2	2	0		
				1	–2	–1	–1	0	
					2	–4	–2	–2	0
v	0	–1	0	6	3	–3	–3	–2	0
n	–2	–1	0	1	2	3	4	5	6

Thus $v[n]$ for $-2 \le n \le 5$ is $[0, -1, 0, 6, 3, -3, -3, -2]$

THE CORRECT ANSWER IS: (D)

18. The function $x(t) = t$ for the time range $0 \le t < 1$ and $h(t) = 1$. The convolution integral is:

$$y(t) = \int_0^t h(\lambda - t)x(\lambda)\,d\lambda = \int_0^t \lambda\,d\lambda = t^2/2$$

THE CORRECT ANSWER IS: (C)

19. $y[k] = \frac{1}{6}(3x[k] + 2x[k-1] + x[k-2])$

Taking the z-transform of both sides yields:

$$Y(z) = \frac{1}{6}\left[3X(z) + 2z^{-1}X(z) + z^{-2}X(z)\right] = \frac{X(z)}{6}\left[3 + 2z^{-1} + z^{-2}\right] = \frac{X(z)}{6}\left[\frac{3z^2 + 2z + 1}{z^2}\right]$$

$$H(z) = \frac{Y(z)}{X(z)} = \frac{1}{6}\left[\frac{3z^2 + 2z + 1}{z^2}\right]$$

THE CORRECT ANSWER IS: (D)

20. With V_1 grounded, the circuit is configured as a non-inverting amplifier and

$$V_o = \left(1 + \frac{R_2}{R_1}\right)V_n$$

where V_n is the voltage applied to the non-inverting terminal of the op amp. V_n can be found by applying the voltage divider rule:

$$V_n = \frac{R_4}{R_3 + R_4}V_2$$

Combining the two results yields:

$$V_o = \left(1 + \frac{R_2}{R_1}\right)\left(\frac{R_4}{R_3 + R_4}\right)V_2$$

$$\frac{V_o}{V_2} = \left(1 + \frac{200\,\text{k}\Omega}{5\,\text{k}\Omega}\right)\left(\frac{200\,\text{k}\Omega}{5\,\text{k}\Omega + 200\,\text{k}\Omega}\right)$$

$$\frac{V_o}{V_2} = 40$$

THE CORRECT ANSWER IS: (C)

21. $v_{DS} = 2$ V

$v_{GS} - V_t = 5 - 1 = 4$ V

$v_{DS} < v_{GS} - V_t$, so the transistor operates in the triode region.

$i_{DRAIN} = 0.5\ [2(4)(2) - 4] = 6$ mA

$$R = \frac{15-2}{6} = 2.17\ \text{k}\Omega$$

THE CORRECT ANSWER IS: (B)

22. Since β is large $V_B \simeq \dfrac{20\ \text{k}\Omega}{100\ \text{k}\Omega} \times 10\ \text{V} = 2\ \text{V}$

$V_E = V_B - V_{BE} = 2 - 0.7 = 1.3$ V

$\therefore\ I_{\text{E}} \simeq I_{\text{C}} = 1$ mA

$V_c = V_{cc} - I_cR_c = 10\ \text{V} - (1\ \text{mA})(4\ \text{k}\Omega) = 6\ \text{V}$

THE CORRECT ANSWER IS: (A)

23. See Solid-State Electronics and Devices in the Electrical and Computer Engineering section of the *FE Reference Handbook.*

THE CORRECT ANWER IS: (A)

24. The average value of the input voltage is:

$$V_{in\ av} = \frac{1}{T}\left[0\left(\frac{T}{3}\right) + 0.1\left(\frac{3T}{6}\right) + (-0.1)\left(\frac{T}{6}\right)\right]$$

$$V_{in\ av} = \frac{0.1}{2} - \frac{0.1}{6} = \frac{1}{3}(0.1)$$

$$\text{Gain} = -\frac{100}{10} = -10$$

$$V_{out\ av} = (-10)\left(\frac{0-1}{3}\right) = -\frac{1}{3} = -0.333\ \text{V}$$

THE CORRECT ANSWER IS: (C)

25. All minterms of the K-map can be covered with two groupings as shown below.

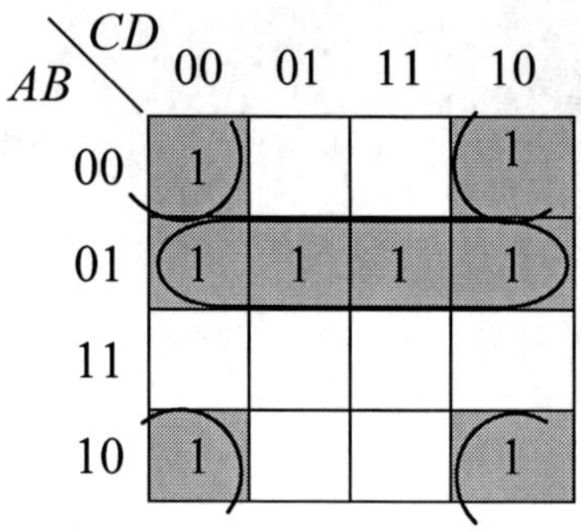

$f(A,B,C,D) = \bar{A}\,B + \bar{B}\,\bar{D}$

THE CORRECT ANSWER IS: (D)

26. If the count is 10, then the D input to the D-type flip is 0, and the JK inputs to the JK flip-flop are both 1, as shown in the figure. After the next CLK signal is applied, the JK flip-flop will toggle from 1 to 0, and the D-type flip-flop will still have a 0 latched in its output. The count will be 00.

THE CORRECT ANSWER IS: (A)

27. $\frac{93}{128}$ can be broken into $\frac{64+16+8+4+1}{128}$ which by decimal is 0.1011101

THE CORRECT ANSWER IS: (B)

28. Indexed addressing uses the contents of some register (R in this problem) as a pointer to the beginning (or end) of a list (array) of values in memory base. An offset (d) is added to the value of the base pointer to determine the location of an operand.

THE CORRECT ANSWER IS: (C)

29. The EPROM will respond whenever CS_4 is asserted, or whenever the address issued by the μp is 100X XXXX XXXX XXXX = 8000 to 9FFF. (X = don't care)

THE CORRECT ANSWER IS: (C)

30. The first loop calculates $\sum_{k=0}^{N-1} S(k)$ and stores this value in X: X = X + S(K)

Next, X is divided by 512, so X = $\dfrac{\sum_{k=0}^{N-1} S(k)}{N}$.

Since X0 = X∗X, X0 = $\left[\dfrac{\sum_{k=0}^{N-1} S(k)}{N}\right]^2$

The second loop calculates $\sum_{k=0}^{N-1} S^2(k)$ and stores this value in X1: X1 = X1 + S(K)∗S(K)

Since X3 = X1/512, X3 = $\dfrac{1}{N}\sum_{k=0}^{N-1} S^2(k)$

Finally, X4 = X3 – X0 = $\dfrac{1}{N}\sum_{k=0}^{N-1} S^2(k) - \left[\dfrac{\sum_{k=0}^{N-1} S(k)}{N}\right]^2$.

THE CORRECT ANSWER IS: (C)

FE Study Material Available for Purchase

FE Supplied-Reference Handbook

Sample Questions and Solutions (in printed and a multidiscipline, CD-ROM format)
are available for the following modules:

Chemical
Civil
Electrical
Environmental
General
Industrial
Mechanical

For more information about these and other Council publications and services, visit us at www.ncees.org or contact our Customer Service Department at (800) 250-3196.